Mathematics Coursework

1-3

Kim Woodward and Chris Cox

John Murray

First published 1989 by
John Murray (Publishers) Ltd
50 Albemarle Street, London W1X 4BD

Book and cover design by Tony Stocks
Cartoons by John Erasmus
Technical illustrations by Technical Art Services
Printed in Great Britain by St Edmundsbury Press, Bury St Edmunds
and bound by Hunter and Foulis, Edinburgh

British Library Cataloguing in Publication Data

Woodward, Kim
Mathematics coursebook 1–3.
1. Mathematics – For schools
I. Title II. Cox, Christopher J.
510

ISBN 0-7195-4565-X

Contents

Introduction

The majority of the teachers that we have spoken to about the coursework element of the GCSE have been concerned, perhaps even a little frightened, particularly in the area of assessing. For many of us, mathematics has been a teaching, doing, marking, traditional subject in which we felt secure and confident. We were proud of our objective examinations, even though we knew that there were times when the results did not reflect the true potential of our pupils. Now, the coursework element seems to shatter our security and leaves us floundering. The idea that *we* have to decide if a piece of work is grade A or grade C, rather than follow a minutely detailed mark-scheme, is alien to most of us. But coursework is here to stay and our pupils depend on us to help them learn the new ways and how to tackle open-ended assignments.

We have been developing this material in an effort to bridge, for both pupils and teachers, the gulf between exercise/answer traditional methods and the open-ended coursework, often mistakenly referred to as 'investigations'. (Investigations are but one facet of GCSE coursework – though possibly the one that traditionalists find hardest to accept.) We have tried to incorporate sufficient structure into each assignment for both pupils and teachers to feel secure and confident, while at the same time the topic opens out, giving scope for the interested motivated pupils to extend themselves to their limits.

In trials, the material has been popular with pupils and, as a result, a suitable marking/assessment/grading scheme has been developed which produces an acceptable correlation between the grades awarded by different markers. Because we wanted the assignments to be suitable for all GCSE abilities, we have devised a grading scheme that gives each pupil the realistic target of obtaining a grade A within each GCSE level.

Because pupils are often working together on the assignments we feel it is necessary to check that each pupil has really learnt from the experience, and has not blindly copied another pupil's ideas. The 'On Your Own' authenticity tests fulfil this function. If the score on the test does not correlate with the grade obtained on the coursework, it will be necessary for the teacher to talk with the pupil concerned to try to discover the reason for the anomaly.

Pupils who have completed these fifteen assignments will be well prepared for the GCSE coursework element.

Elements of the package

Target Group	Stage One: pupils aged 11 to 12 ('Year One') Stage Two: pupils aged 12 to 13 ('Year Two') Stage Three: pupils aged 13 to 14 ('Year Three')
Topics	One at each stage on each of: Practical geometry Investigation Everyday application Problem-solving Statistics/probability
Teachers' Material	Teaching notes Answer and marking guide Grading scheme
Pupils' Material	'On Your Own' authenticity test Instruction sheet Worksheet

Using the material

Although written for users of *Understanding Mathematics* (*UM*) and *Steps in Understanding Mathematics* (*SUM*), the assignments are equally usable with any other scheme. We have suggested in the contents list (page 3) where each assignment best fits into the schemes, sometimes as an introduction to a topic, sometimes as an extension.

Pupils enjoy working in small groups, though they should be encouraged to write their reports independently. At all times, teachers should be involved in discussing what the pupils are doing, guiding but not leading, hinting but not directing, asking questions but not answering them. We aim to make pupils more dependent on their own resources, not slavish followers of instructions. The 'On Your Own' tests should identify pupils who have produced work that is not based on their own understanding, and counselling is then required.

Marking the coursework

In our experience within schools and on courses, marking methods that depend on 'a feel for the submitted work' lead to an unacceptably wide variation in grades. We found grades varying from B to E on the same pieces of coursework; and when children's future careers may well hang on whether they obtain a C or a D grade at GCSE this is quite unacceptable. Although we applaud the move away from question/answer traditionalism, we must nevertheless ensure that the GCSE grades awarded do truly reflect the ability of the candidates at least as well as they have done in the past.

Our marking scheme may seem rather restrictive to those in favour of a subjective approach. It may be ignored of course! However, we have found it highly successful in trials, leading to a close correlation of ability and grades obtained and thus to fair grading; it is always possible to adjust the grade to reflect an outstanding effort in one area of the assignment.

We would welcome your observations after using this material, particularly suggestions for its improvement.

Kim Woodward
Chris Cox

Criteria Referencing

Each piece of coursework has been assessed for its mathematical content, based on the following criteria. The authors thank Mr R. W. Strong, the Mathematics Advisor for Somerset, and the Somerset Mathematics Advisory Teachers for permission to use this model.

The codings given are used in the Matrix on page 7.

Starting

Evidence of:

S1 Understanding the problem.
S2 Devising a strategy for solution.
S3 Collecting relevant resources.
S4 Organising the information and resources.

Doing

Evidence of:

D1	Measuring	Using correct units.
D2		Using appropriate instruments correctly.
D3		Estimating quantities.
D4	Calculating	Understanding of number.
D5		Approximating numerical solutions reasonably.
D6		Performing calculations correctly.
D7	Representing	Using diagrams to give information.
D8		Using diagrams to extract information.
D9		Interpreting scales.
D10		Making a scale representation.
D11		Interpreting graphical information.
D12		Presenting information in graphical format.

Reviewing

R1 Checking results.
R2 Suggesting what might happen if.
R3 Generalising from particular outcomes.
R4 Justifying and proving outcomes.
R5 Communicating within/about the activity.

A few notes on the use of these criteria in the Somerset Model

Each piece of coursework is looked at and a decision reached as to which of the above criteria apply to it. Each student's work is then examined for each of the expected criteria, under the headings 'much', 'some' and 'no'. 'Much' and 'some' are further divided into upper and lower levels. From this evidence a mark is given within each of the three main sections, Starting, Doing and Reviewing, and thus a total score for the work is obtained.

	S				D												R				
	1	2	3	4	1	2	3	4	5	6	7	8	9	10	11	12	1	2	3	4	5
Spirolaterals	*			*	*						*	*					*	*	*	*	*
Sinbad	*	*									*	*					*	*	*	*	*
All Booked Up	*	*		*			*		*	*	*	*					*	*		*	*
Martin's Bedroom	*	*		*	*	*		*		*	*		*				*				*
I've Won	*		*	*	*	*			*	*	*	*	*	*	*	*	*		*		*
Countdown	*	*						*		*	*	*					*	*	*	*	*
Fruity	*		*				*	*	*	*							*	*	*	*	*
Speed Limits	*	*		*	*	*	*	*	*	*	*	*	*	*	*		*		*	*	*
Nets	*	*	*	*	*	*				*							*		*	*	*
Hold the Line	*	*	*	*	*	*	*	*		*	*	*	*				*				*
A Day Out	*	*	*	*	*		*			*	*	*	*	*	*	*	*		*		*
Double-Glazing	*	*	*		*	*	*	*		*	*	*					*				*
Get Fit	*			*	*	*	*	*	*	*				*	*	*	*		*	*	*
Cheese Snacks	*	*	*	*	*	*	*	*	*	*	*						*		*	*	*
Mirror, Mirror	*	*		*							*	*					*	*	*		*

Spirolaterals

Topic	Practical geometry.
Content	LOGO-type patterns. Ideas of turning, 90°, 60°, etc.
Assumed knowledge	None.
Integration	UM1 and SUM1: to precede *Angles: the protractor; kinds of angle.*
Administration	Individual or small groups. Coloured pencils make the results more attractive. Encourage pupils to try their own number sequences. They should write the sequence used by each drawing. Where possible pupils should use a computer LOGO program.

Answers and Marking Guide

Step	Notes	Max. Marks
1	Correct drawing	1
2	Correct drawing, see Figure T1	2
3	(a) Correct drawing, see Figure T2	2
	Correct drawing, see Figure T3	2

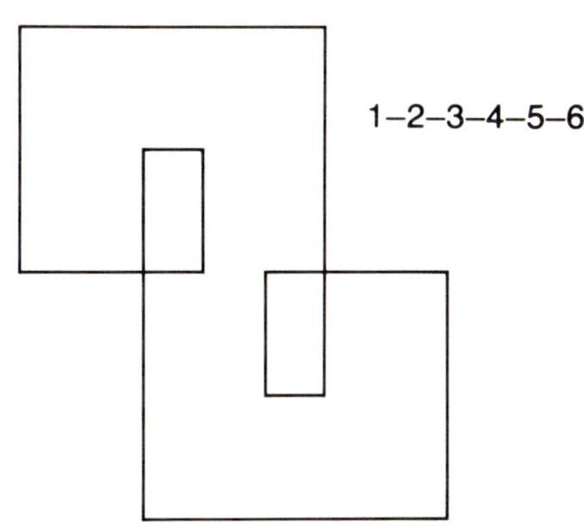

Figure T1

1–2–1–2

Figure T2

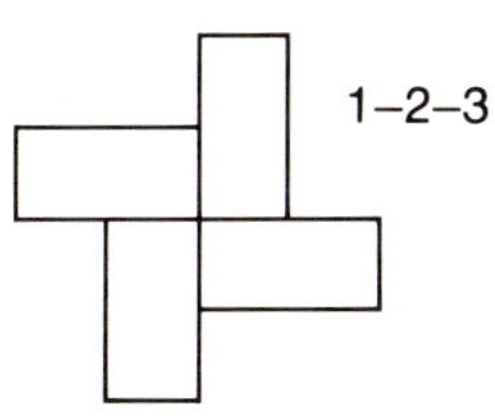

Figure T3

Max. Marks

(b) Correct drawing, see Figure T4 — 1

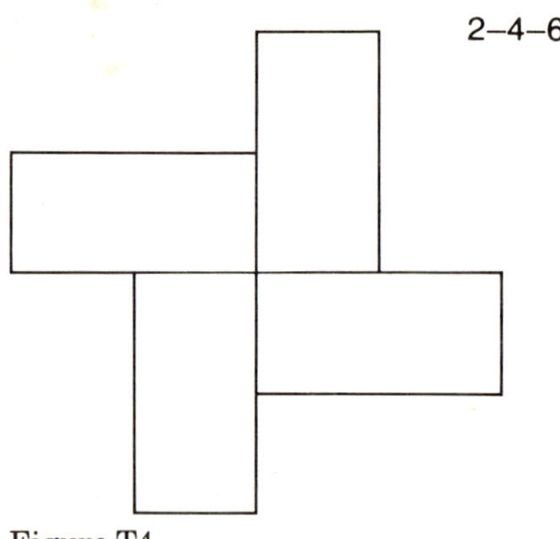

Figure T4

		Max. Marks
	Saying that lengths are doubled	1
	Saying that area is four times as big	1
	(c) Clear description, e.g. mirror images	2
4	Accurately drawn and labelled drawing	1

5

1	4
1–2	2
1–2–3	4
1–2–3–4	never
1–2–3–4–5	4
1–2–3–4–5–6	2
1–2–3–4–5–6–7	4
1–2–3–4–5–6–7–8	never

		Max. Marks
	Pupils may draw as many as they like. However the spirolaterals 1–2–3–4 and 1–2–3–4–5–6–7–8 spiral off the paper and never return to the start. *Note:* Results in the chart not only hold for the given sequences but also for any sequence with the same number of terms. For each answer award a half mark	3
6/7/8	It is not expected that all these extensions will be attempted. Credit should be given for LOGO work	4
		20

Suggested Grading Scheme

	A	B	C	D	E
Level 3	19+	17–18	14–16	12–13	10–11
Level 2	17+	14–16	12–13	10–11	8–9
Level 1	14+	12–13	10–11	8–9	6–7

On Your Own Answers

		Max. Marks
1	Correct diagram, see Figure T5	4
2	3–5–7 has to be repeated four times	2
3	You will get a mirror image	4
		10

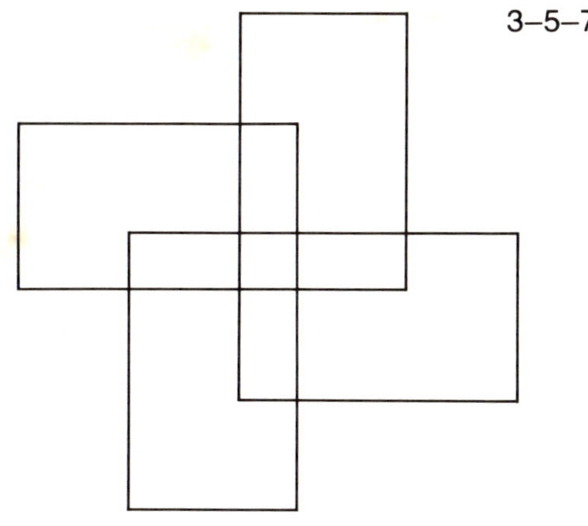

Figure T5

Spirolaterals

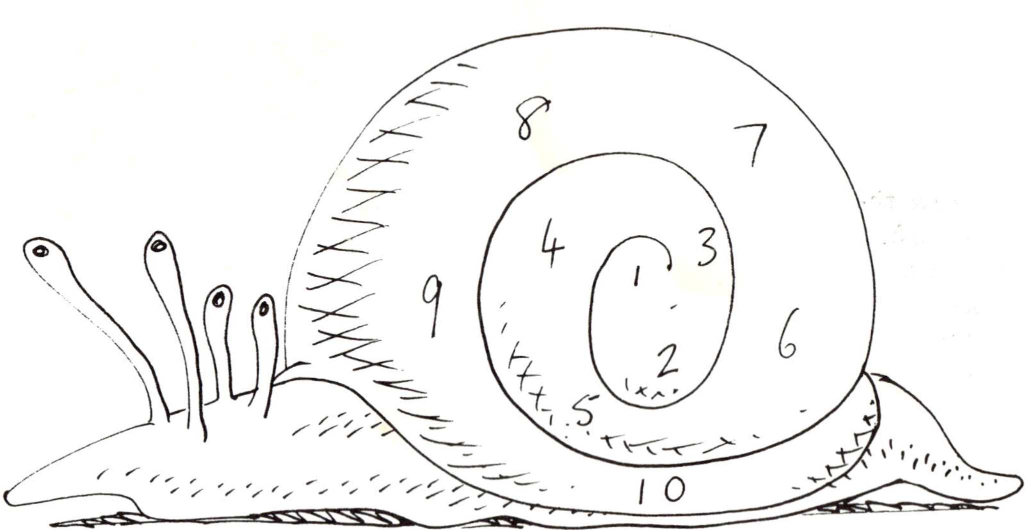

1 Look at the snail in Figure 1. The snail can be made into a spirolateral by giving each number a length, one square for a 1, two squares for a 2, three squares for a 3, etc. Turn right after drawing each length, as in Figure 2. Draw Figure 2 on squared paper and repeat the sequence four times. The spirolateral should look like Figure 3.

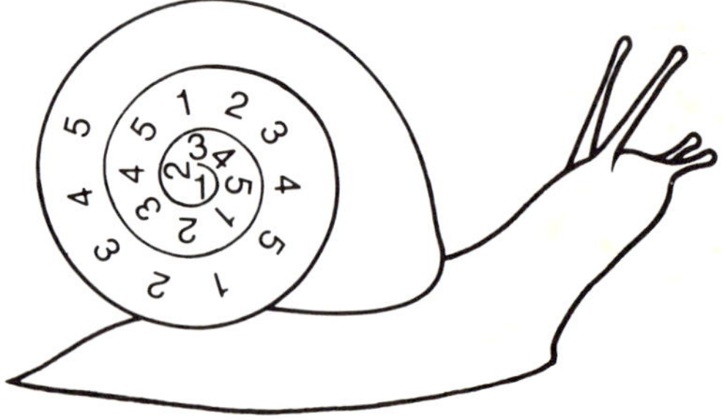

Figure 1

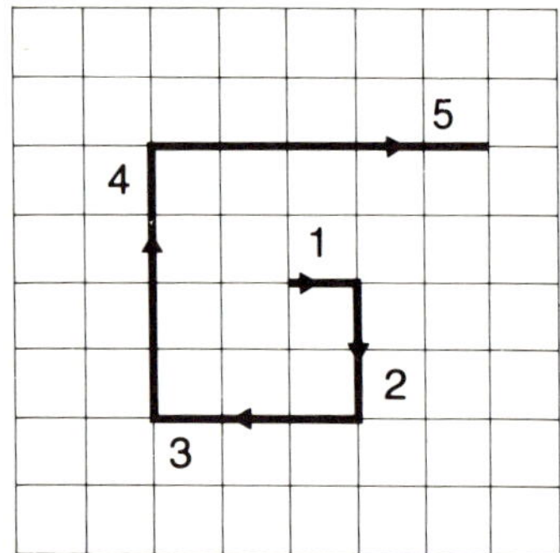

Figure 2

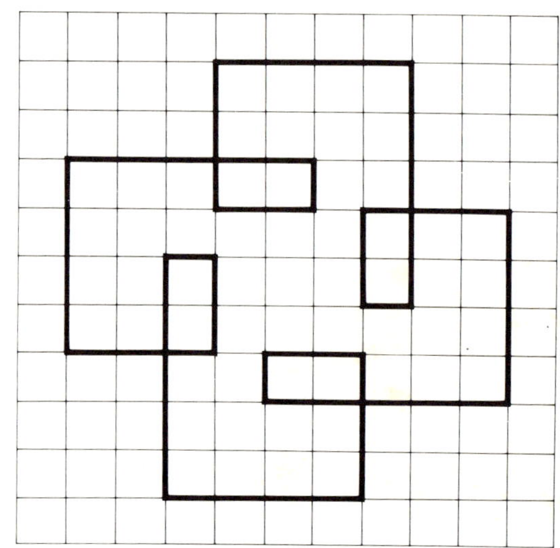

Figure 3

2 Figure 4 is a new snail for you to make into a spirolateral.

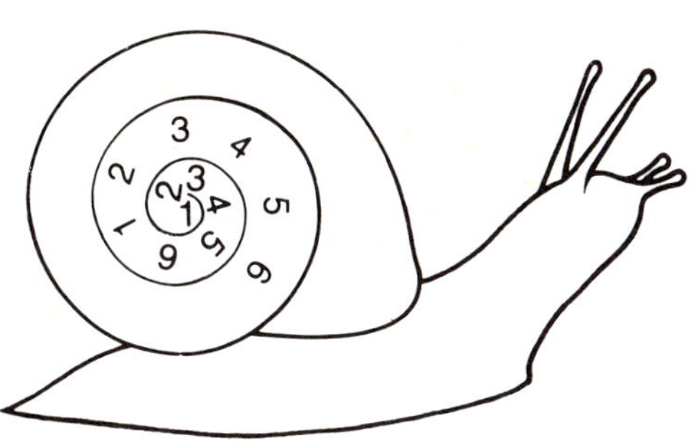

Figure 4

3 If we draw the spirolateral 1–1–1–1 it looks like Figure 5.

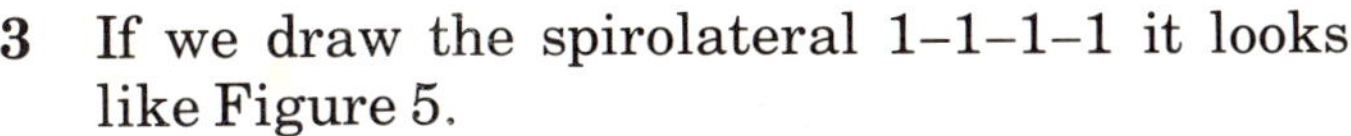

(a) Now draw the spirolaterals 1–2–1–2 and 1–2–3–1–2–3–1–2–3–1–2–3 on squared paper.

(b) Draw the spirolateral for 2–4–6 and compare it with the 1–2–3 spirolateral.

(c) How does 3–4–5 with right-turns compare with 3–4–5 using left-turns?

Figure 5

4 Now try a spirolateral using your phone number or your date of birth. (Use each figure separately.) Figure 6 shows an example.

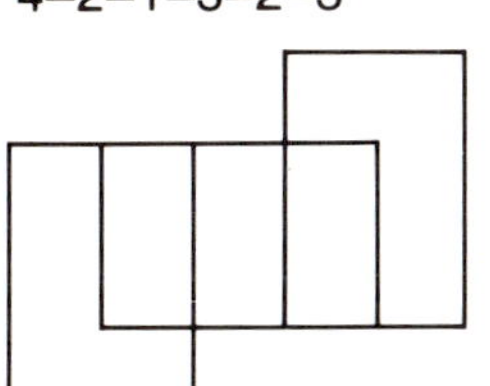

Figure 6

5 Copy table 1 and try to fill in the missing numbers. Is there a pattern to your results?

Number sequence	How many times do you need to go through the numbers before you get back to the start?
1	4
1–2	
1–2–3	
1–2–3–4	
1–2–3–4–5	4
1–2–3–4–5–6	
1–2–3–4–5–6–7	
1–2–3–4–5–6–7–8	

Table 1

6 Try 60° turns, using isometric paper. Which ones go back to the beginning?

7 Mix right and left turns. For example 1–2–3–4–5–6 with all right turns, except left turns on 3.

8 Try using several terms of the Fibonacci sequence, 1, 1, 2, 3, 5, 8, etc.

Spirolaterals

ON YOUR OWN

1 Draw the spirolateral 3–5–7 with right-hand turns.

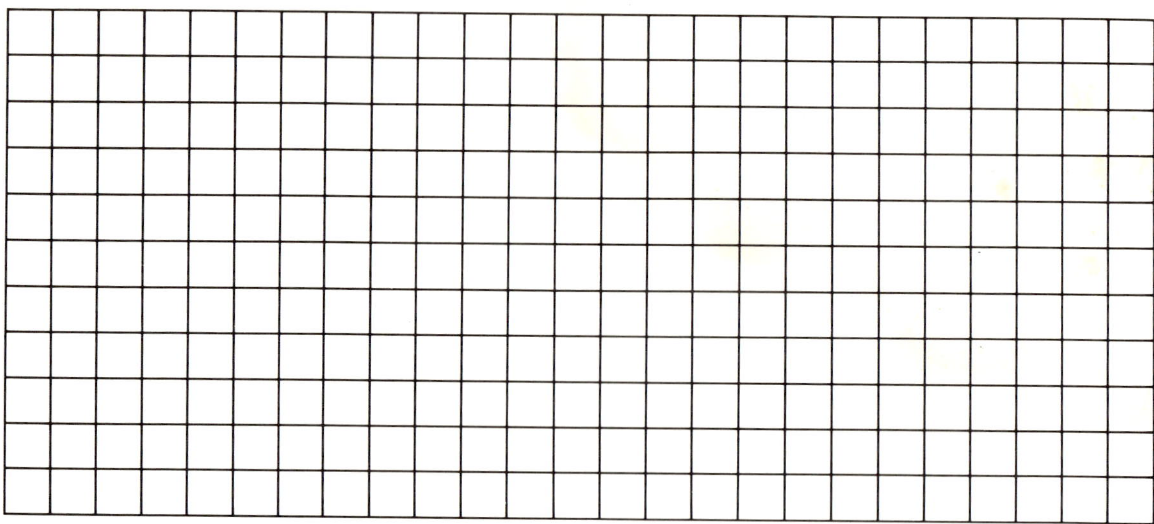

2 How many times do you have to repeat the sequence 3–5–7 before it reaches the beginning again?

3 What happens if you draw the spirolateral 3–5–7 with left turns instead of right turns? Compare the two patterns.

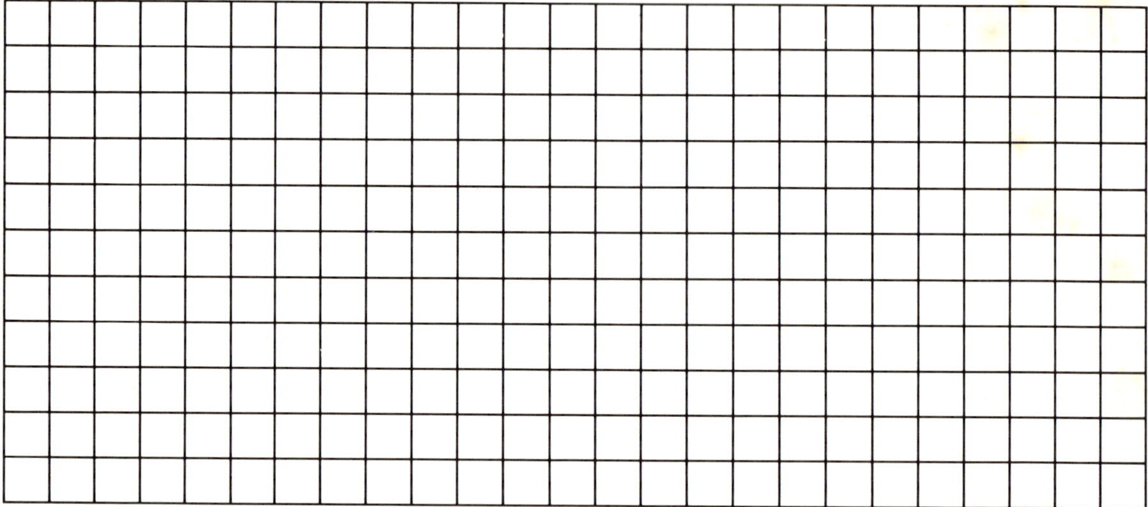

Sinbad

Topic	Investigation.
Content	Investigation of routes. Position by co-ordinates.
Assumed knowledge	None.
Integration	UM1 and SUM1: to precede *Graphs: co-ordinates*.
Administration	Pairs. It is essential that pupils read through all the information before tackling the work.

Answers and Marking Guide

Step	Notes	Max. Marks
1	Evidence of understanding the rules and playing some games correctly	2
2	The Old Man can win in eight ways, see Figures T1, T2 and T3	
	One correct solution	2
	Alternative solutions: one	1
	over one	1

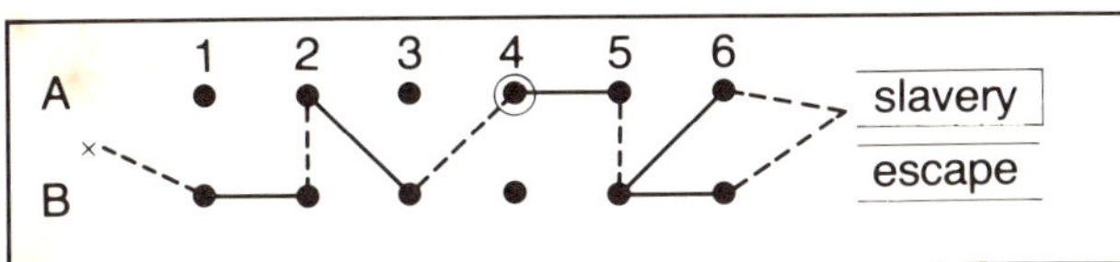

Figure T1

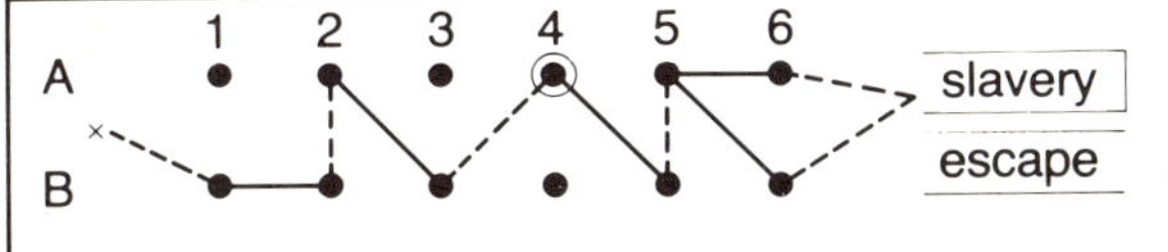

Figure T2

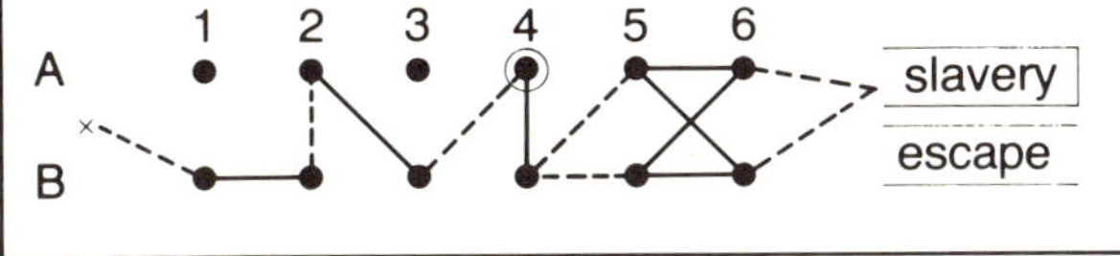

Figure T3

Step	Notes	Max. Marks
3	Sinbad will win if he crosses over A4/B4 and A5/B5 (see Figure T4)	
	Searching for strategies	2
	Correct solution	3

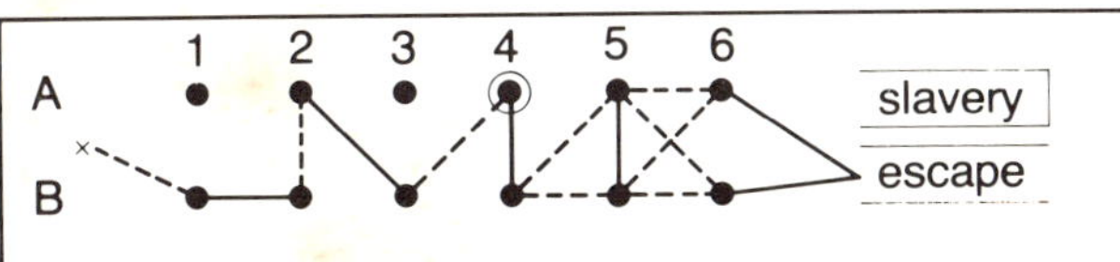

Figure T4

Max. Marks

4	Sinbad will always win if he crosses straight over (see Figure T5)	
	Searching for strategies	2
	Correct solution	3

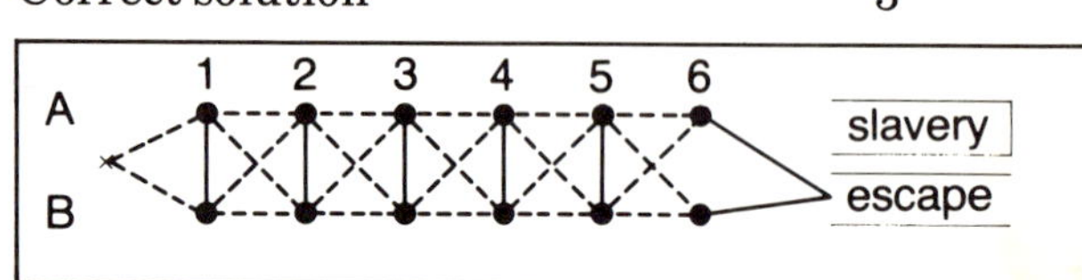

Figure T5

5	The strategy of step 4 works for any number of points	2

6 If Sinbad reaches A4 cr C4 first he can force a win. He must avoid choosing A3, B3 or C3, otherwise the Old Man can reach A4/C4 first. He must head for A2 or C2 to be sure of making the Old Man go to A3, B3 or C3, however hard he tries to avoid this. Figures T6 and T7 show possible end moves. Figures T8, T9 and T10 show possible starts.

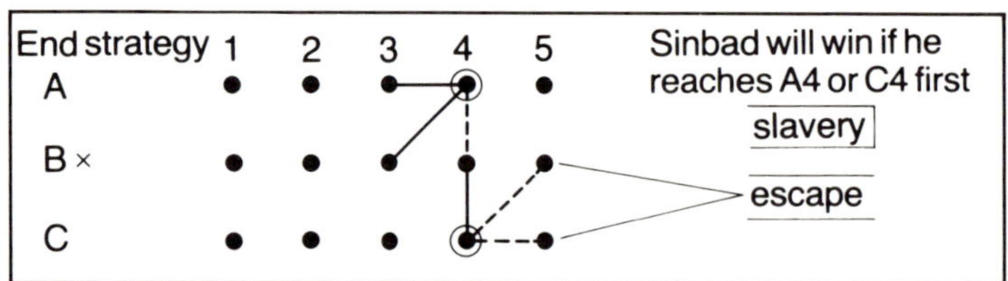

Figure T6

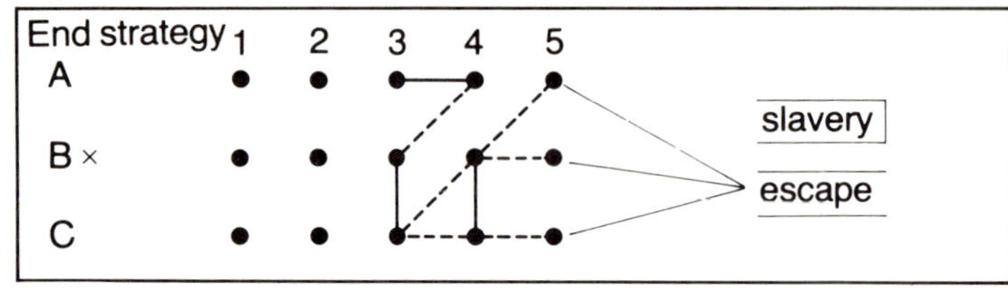

Figure T7

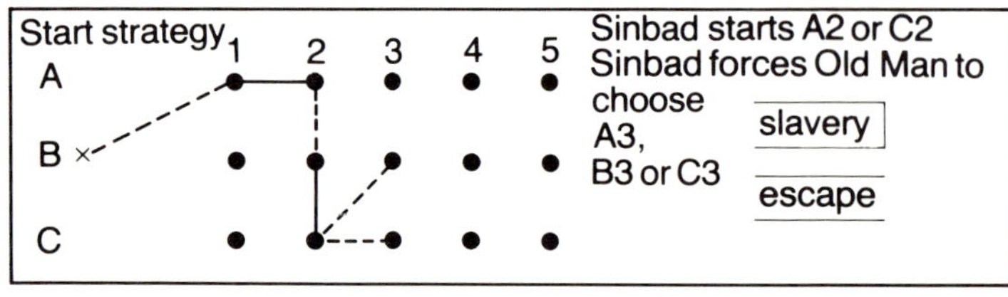

Figure T8

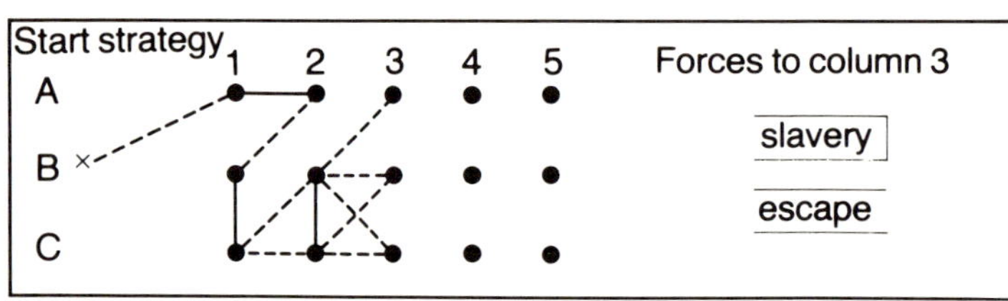

Figure T9

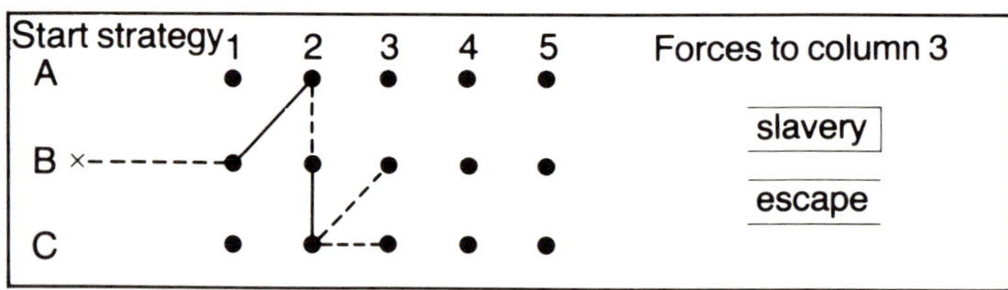

Figure T10

		Max. Marks
	Any successful crossing	3
	Realising the importance of points in columns 2, 3 and 4	3
	Strategies to force the Old Man to choose A3, B3 or C3	3
7	Attempts to develop the idea	3
		30

Suggested Grading Scheme

	A	B	C	D	E
Level 3	26+	23–25	18–22	15–17	12–14
Level 2	22+	18–21	14–17	11–13	9–10
Level 1	18+	15–17	11–14	8–10	6–7

On Your Own Answers

		Max. Marks
1	The next cross should be as in Figure T11	2
2	Saying that it will be a draw	2
	Showing the possible endings (see Figure T12)	2+2
		8

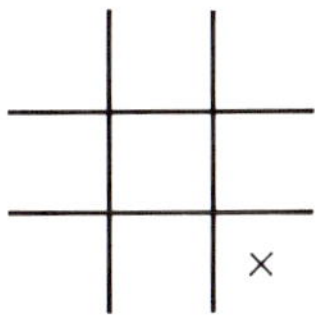

Figure T11

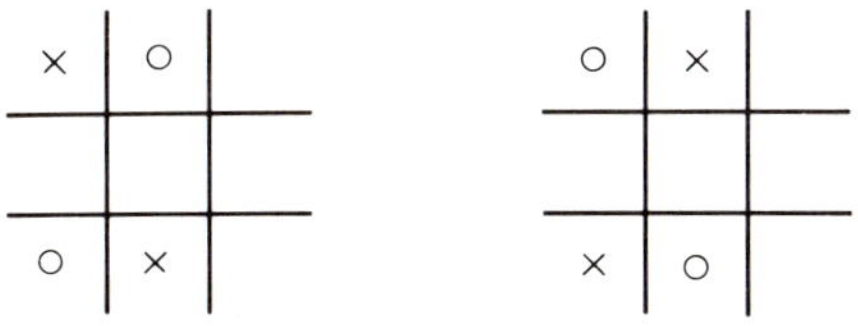

Figure T12

Sinbad

Sinbad has the Old Man of the Sea on his back. Sinbad can escape if he can cross the sea using the turtles' backs as stepping stones. Worksheet One shows the turtles as dots. The Old Man of the Sea chooses the first turtle, A1 or B1. Then Sinbad is allowed to choose. They take it in turns to choose until Sinbad reaches A6 or B6. If the next choice is Sinbad's then he can choose the escape route, where the Roc bird plucks the Old Man from Sinbad's back. But if it is the Old Man's turn to choose he can choose to keep Sinbad as his slave for ever.

RULES

1 Sinbad can step on a turtle in front, behind, to his left or right, or diagonally across a square. Figure 1 shows the possible moves from turtle A4.

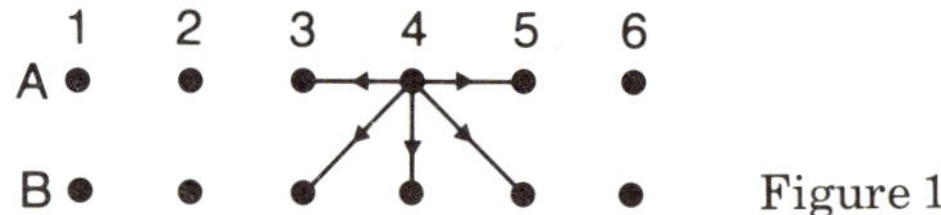

Figure 1

2 A turtle may be stepped on only once.

3 If there is no turtle left to choose to step on then they are both captured by the evil Roc bird.

4 The Old Man of the Sea insists that he chooses the first turtle.

5 Sinbad is not allowed to recross his track.

Figure 2 shows one possible crossing, which results in Sinbad being made a slave. Sinbad's choices are shown with a straight line, the Old Man's choices with a dashed line.

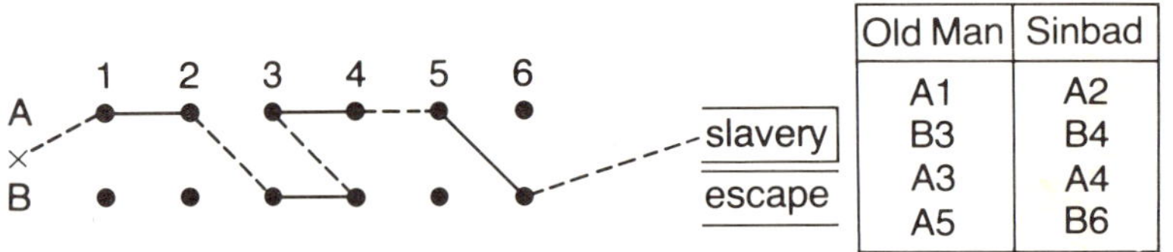

Old Man	Sinbad
A1	A2
B3	B4
A3	A4
A5	B6

Figure 2

1 Try out a few crossings on Worksheet One. Record your moves in the boxes provided.

2 Table 1 shows the start of one crossing. Can the Old Man of the Sea win from this point? If so, how?

Old Man	Sinbad
B1	B2
A2	B3
A4	

Table 1

3 From the same start given in step 2, can Sinbad choose turtles so that he is certain of escaping, however hard the Old Man of the Sea tries to stop him? If so, how?

4 Is there a plan that Sinbad can use so that he is always certain to escape if the Old Man has the first choice? If so, try to explain the plan clearly.

5 What difference would it make to Sinbad's escape plan if there were more pairs of turtles, say seven pairs, or eight pairs?

6 Worksheet Two shows a different set of turtles. Using the same rules can you find a plan for Sinbad so that he always escapes on this crossing, no matter how hard the Old Man of the Sea tries to stop him?

7 Investigate further. You could change the pattern again, or change the rules.

Sinbad

WORKSHEET ONE

	1	2	3	4	5	6	
A	●	●	●	●	●	●	slavery
×							
B	●	●	●	●	●	●	escape

Old Man	Sinbad

	1	2	3	4	5	6	
A	●	●	●	●	●	●	slavery
×							
B	●	●	●	●	●	●	escape

Old Man	Sinbad

	1	2	3	4	5	6	
A	●	●	●	●	●	●	slavery
×							
B	●	●	●	●	●	●	escape

Old Man	Sinbad

	1	2	3	4	5	6	
A	●	●	●	●	●	●	slavery
×							
B	●	●	●	●	●	●	escape

Old Man	Sinbad

	1	2	3	4	5	6	
A	●	●	●	●	●	●	slavery
×							
B	●	●	●	●	●	●	escape

Old Man	Sinbad

	1	2	3	4	5	6	
A	●	●	●	●	●	●	slavery
×							
B	●	●	●	●	●	●	escape

Old Man	Sinbad

WORKSHEET TWO

	1	2	3	4	5	
A	•	•	•	•	•	
						slavery
B ×	•	•	•	•	•	
						escape
C	•	•	•	•	•	

	1	2	3	4	5	
A	•	•	•	•	•	
						slavery
B ×	•	•	•	•	•	
						escape
C	•	•	•	•	•	

	1	2	3	4	5	
A	•	•	•	•	•	
						slavery
B ×	•	•	•	•	•	
						escape
C	•	•	•	•	•	

	1	2	3	4	5	
A	•	•	•	•	•	
						slavery
B ×	•	•	•	•	•	
						escape
C	•	•	•	•	•	

	1	2	3	4	5	
A	•	•	•	•	•	
						slavery
B ×	•	•	•	•	•	
						escape
C	•	•	•	•	•	

Sinbad

ON YOUR OWN

1 Look at the game of noughts and crosses in Figure 1. Which position should the cross go in next in order to win?

2 In the game in Figure 2 what is going to happen if each player tries to stop the other from winning?

Show possible endings to this game.

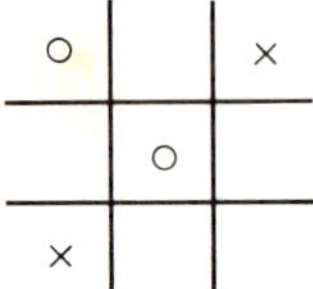

Figure 1

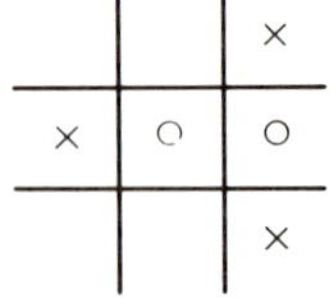

Figure 2

All Booked Up

Topic	Problem-solving.
Content	Organisation of a timetable. Estimation and costing.
Assumed knowledge	12-hour-clock time. Costing: £ per hour. Approximation to nearest unit.
Integration	UM1 and SUM1: to precede *Time: clocks.*
Administration	Individual or small groups. With less able pupils, some discussion of when in the evening the events should take place, and the implications of the charges and costs.

Answers and Marking Guide

Step	Notes	Max. Marks
1	One working timetable, even if timings are not ideal	5
	Sensible timings	5
	Good reasons	4
	Alternative timetables	2+2
2	Calculation of total cost: £182·00	2
	Calculation of total hours: $21\frac{1}{2}$	2
	Method to calculate charge per hour	2
	Sensible charge (say £8·50 to £9·00 per hour)	2
3	Bills: Badminton × 4 Aerobics × 2 Steel Band × $2\frac{1}{2}$ Gym Club × $4\frac{1}{2}$ Drama Society Dinner × 5 History Society × $1\frac{1}{2}$ Disco × 2 −1 each error or omission	4
		30

Suggested Grading Scheme

	A	B	C	D	E
Level 3	26+	22–25	18–21	14–17	12–13
Level 2	21+	18–20	14–17	12–13	10–11
Level 1	18+	15–17	12–14	10–11	8–9

On Your Own Answers

Emma's evening should look something like this:

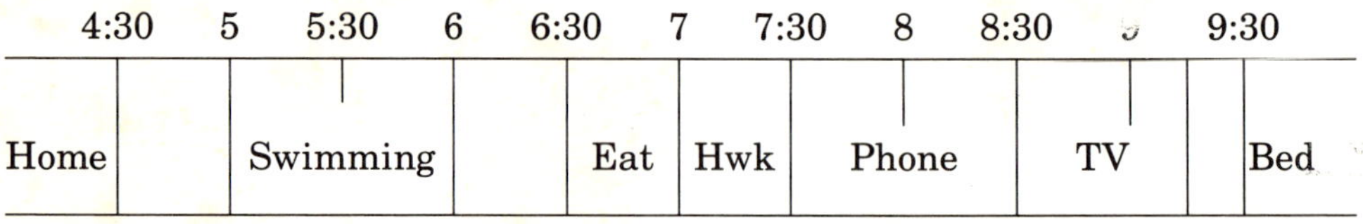

4:30	5	5:30	6	6:30	7	7:30	8	8:30	9	9:30

Home		Swimming		Eat	Hwk	Phone	TV		Bed

	Max. Marks
Correct order (−1 each mistake)	5
Organisation of answer	3
	8

All Booked Up

Sarah is in charge of organising the bookings for the local community-centre hall. During the day the hall is used by the local secondary school, but after 4 p.m. other groups may book the hall and its equipment and facilities.

Here is a list of bookings for the hall for the week:

Badminton (adult group)	Two 2-hour sessions
Aerobics (adult group)	One 2-hour session
Steel Band rehearsal	One $2\frac{1}{2}$-hour session
School play rehearsal	One 3-hour session
Gym Club (5–11 year olds)	Three $1\frac{1}{2}$-hour sessions
Drama Society Dinner	Friday 7–12 p.m.
Local History Society	One $1\frac{1}{2}$-hour session
Under 14's disco	7–9 p.m.

1 Try to fit these groups into the Booking Diary on the Worksheet. Give reasons for your choices. Are there any other sensible ways to book all the groups in? If so, write out not more than two other ways.

2 In order for the community hall to pay for itself Sarah has to charge each of the groups for the amount of the time they use the hall. Here is a list of her costs for an average week:

Gas	£75·00
Electricity	£35·00
Cleaning	$7\frac{1}{2}$ hours at £4·25 per hour
Sundries	£15·00

She must also allow £25·00 per week extra to buy and repair equipment. Work out how much she should charge per hour.

3 Write out the bills for each of the groups using the hall. The school may use the facilities between 9 a.m. and 4 p.m. free of charge and rehearsals for the school play are also free of charge.

All Booked Up

WORKSHEET

COMMUNITY HALL BOOKINGS

	4 p.m.	5 p.m.	6 p.m.	7 p.m.	8 p.m.	9 p.m.	10 p.m.
Monday							
Tuesday							
Wednesday							
Thursday							
Friday							

COMMUNITY HALL BOOKINGS

	4 p.m.	5 p.m.	6 p.m.	7 p.m.	8 p.m.	9 p.m.	10 p.m.
Monday							
Tuesday							
Wednesday							
Thursday							
Friday							

All Booked Up

ON YOUR OWN

Emma has a busy evening ahead. She has to fit in the following activities:

English homework	30 minutes
Swimming Club	5–6 p.m.*
Eat	6:30 ($\frac{1}{2}$ an hour)
Home from school	4:30 p.m.
Watch her favourite TV programme	8:30 p.m. (45 mins)
Bed	9:30 p.m.
Expecting a phone call from her aunt in Australia	About 8 p.m.

*Allow 15 min travelling time each way.

Draw up a time chart to show how she can best fit all these things in.

Martin's Bedroom

Topic	Everyday application.
Content	Practical room planning. Combinations of lengths to fill space. Nets and model making.
Assumed knowledge	Metric lengths of mm, cm and m. Simple scales for drawings.
Integration	UM1 and SUM1: to follow *Scales: scale drawing.*
Administration	Individual. Worksheet Two should only be given to pupils unable to devise their own method of organising the answer. It should then be noted that this worksheet has been provided.

Answers and Marking Guide

Step	Notes	Max. Marks
1	Figure T1 shows the most likely solution. Any satisfactory solution should be accepted	4
2	The alcove is 9 cm scaled, 1800 mm true. This can be filled with: 2 singles and 1 double 4 singles 2 doubles	4
3	(see table below)	

Width	No. @900 mm	No. @450 mm
450 mm	0	1
900 mm	1	0
	0	2
1350 mm	1	1
	0	3
1800 mm	2	0
	1	2
	0	4
2250 mm	2	1
	1	3
	0	5
2700 mm	3	0
	2	2
	1	4
	0	6

Correct answers	4
Organisation	2

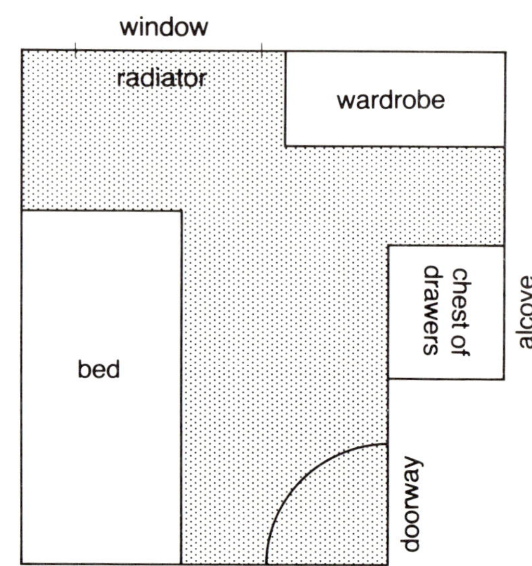

Figure T1

		Max. Marks
4	Any sensible answer. (At the time of writing, MFI offer 300, 500, 600, 1000, and 1200 mm sizes.)	2
5	Sensible kit	2
	Drawings; model; etc.	2
		20

Suggested Grading Scheme

	A	B	C	D	E
Level 3	18+	16–17	12–15	10–11	8–9
Level 2	16+	14–15	11–13	9–10	7–8
Level 1	13+	11–12	9–10	7–8	5–6

On Your Own Answers

		Max. Marks
1	11 cm	2
2	2200 mm	2
3	Sizes of cupboards	

1000 mm	600 mm	500 mm	300 mm
1	2		
1	1		2
1			4
	2	2	
	1	2	2
		2	4

	Max. Marks
(–1 each wrong or missing)	4
Presentation	2
	10

Martin's Bedroom

1 On Worksheet One is a plan of Martin's bedroom and the furniture he wants to put in it. Cut out the furniture. Arrange and stick down the furniture in the bedroom. Bear these things in mind:

(a) The doorway must not be blocked.
(b) The radiator below the window should not have any furniture in front of it, as this means the room will not be properly heated.
(c) You must allow half a metre in front of the wardrobe and chest of drawers in order to be able to open them.
(d) Martin likes one side of his bed against a wall.

2 Martin persuades his parents to buy him a new set of cupboards instead of the wardrobe and the chest. The cupboards are available in two widths, 450 mm and 900 mm (Figure 1). Martin decides to fit these in the alcove by the door. What different sets of cupboards could they buy to exactly fit the alcove?

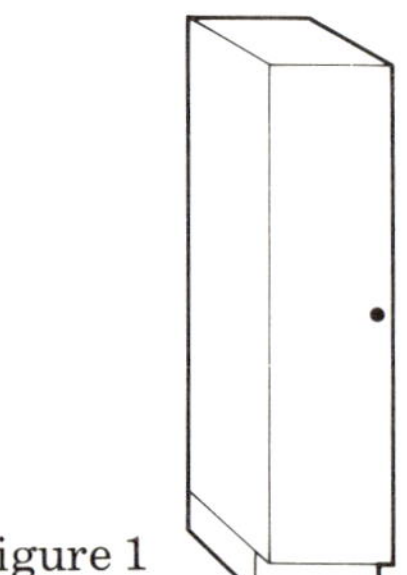

Figure 1

3 The local DIY store prepares information for people who want to fit any space up to 2700 mm. They work out all the different widths that could be fitted exactly. Set out this information if the two cupboards used are the same size as those given in step 2.

4 Older houses have rooms that are usually multiples of feet. One foot is about 300 mm. What metric size cupboards would be best for such houses? If you can, compare your answer with the sizes offered by a retailer such as MFI.

5 Some cupboards are supplied as a kit which has to be fitted together at home. Decide what parts should be in one of these kits, giving sizes and quantities. You may like to sketch your cupboard or make a model of it.

Martin's Bedroom WORKSHEET ONE

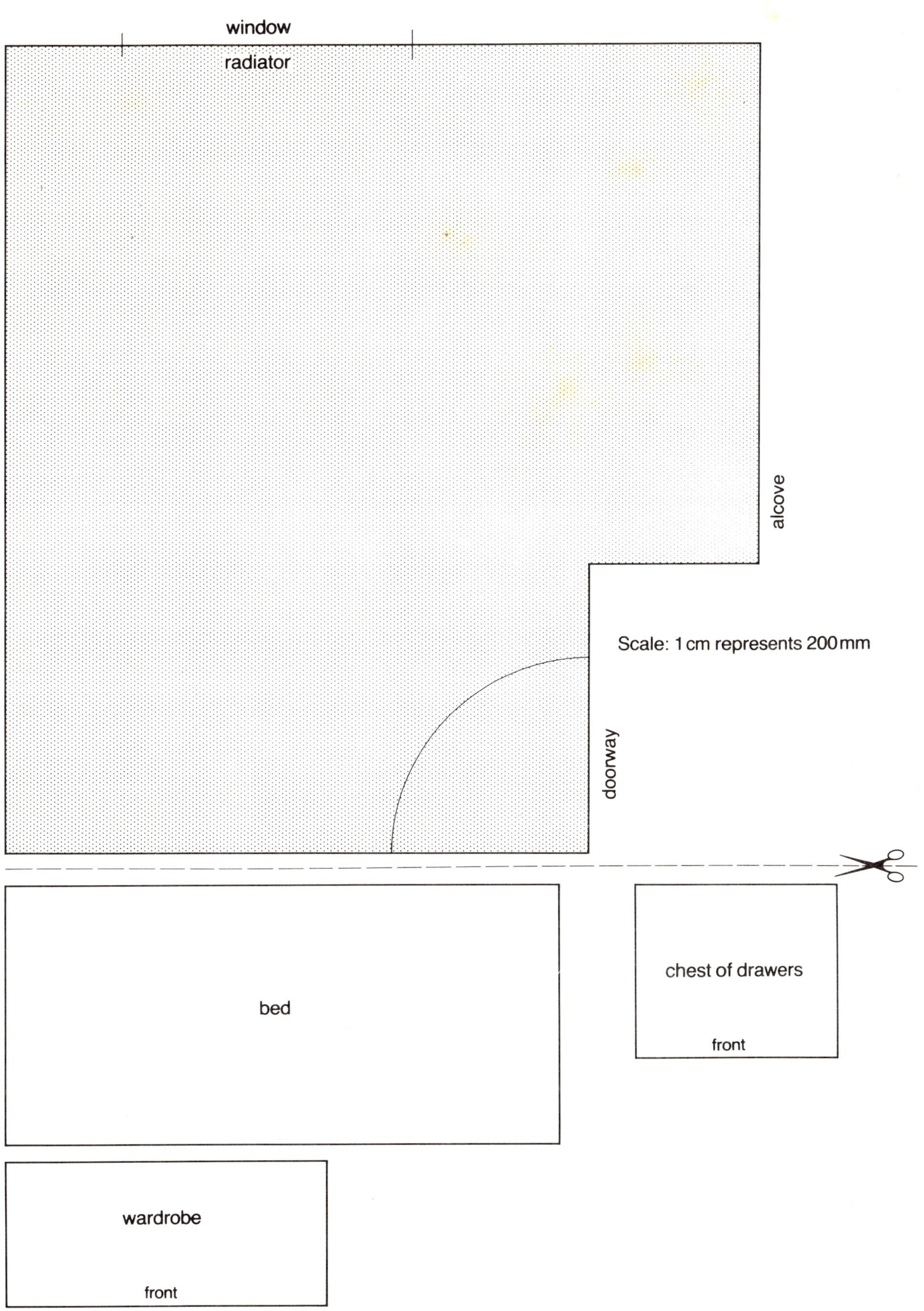

Martin's Bedroom WORKSHEET TWO

Here is a table that might help with question 3.

Total width	**How many cupboards?**	
	900 mm	**450 mm**
450 mm	0	1
900 mm		
1350 mm		
1800 mm		
2250 mm		
2700 mm		

Martin's Bedroom

ON YOUR OWN

1 Figure 1 is a plan of a kitchen. Measure the distance marked 'x' on the plan.

Figure 1

2 How long would this wall be in the actual kitchen in millimetres?

3 The kitchen wall units available from a local DIY store are shown in Figure 2. Choose a range of units that would fit exactly along the wall you have measured.

Find all the different combinations of units that will fit exactly. Set your answer out as clearly as you can.

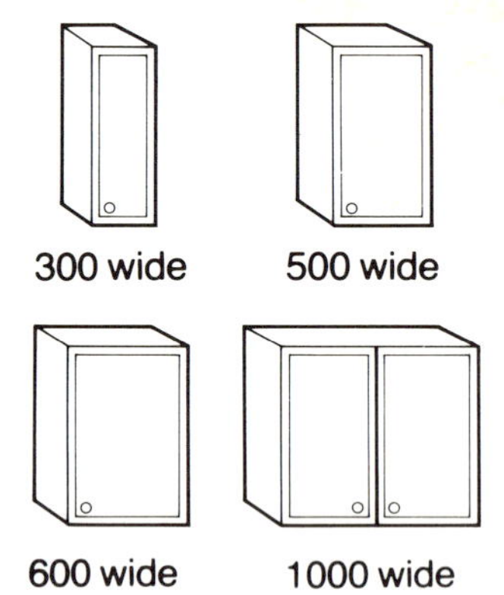

Figure 2

I've Won

Topic	Statistics.
Content	Illustration of statistical information. Conducting a survey. Interpreting results of a survey.
Assumed knowledge	Bar-graph and pie-chart construction.
Integration	UM1 and SUM1: to follow *Statistics: charts*.
Administration	Groups of three or four.

Answers and Marking Guide

Step	Notes	Max. Marks
1	Bar-chart (see Figure T1)	3
	Pie-chart (see Figure T2)	3
2	Programmes should be in categories. It is not necessary to write them all out. A tally chart is sensible	2
3/4	Sensible, clear survey questions	2
5	Clear, organised results table	2
6	Data collection	2
7	Presentation of data	3
8	Newspaper report	3
		20

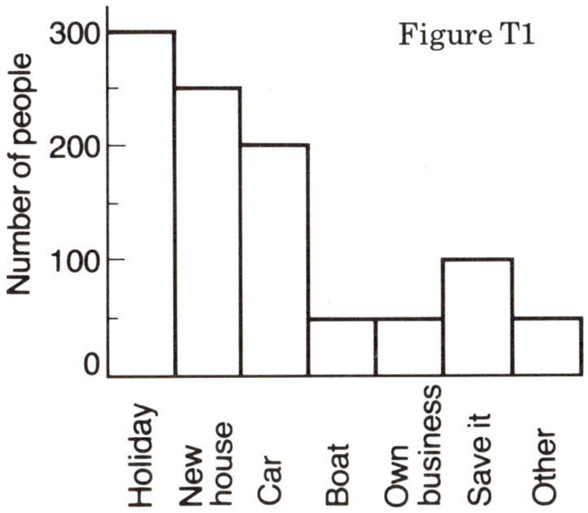

Figure T1

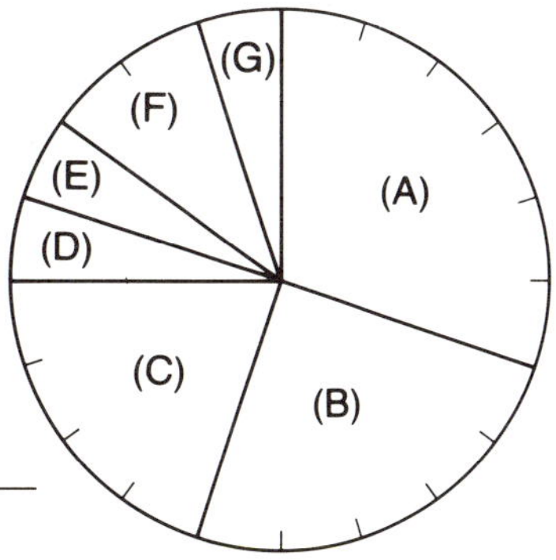

Figure T2

Key
A: Holiday
B: New house
C: Car
D: Boat
E: Own business
F: Save it
G: Other

Suggested Grading Scheme

	A	B	C	D	E
Level 3	19+	17–18	14–16	12–13	10–11
Level 2	17+	14–16	12–13	10–11	8–9
Level 1	14+	12–13	10–11	8–9	6–7

On Your Own Answers

		Max. Marks
1	(a) 22 pupils	2
	(b) 57%	2
	(c) Week 3	2
2	A line graph	2
3	Answer (c): 100–1000	2
		10

I've Won!

What is the first thing you would do with the money if you won a fortune? In our survey we asked one thousand people of all ages and backgrounds. Figure 1 shows the results as a pictogram.

1 From the information in Figure 1, fill in the pie-chart and the bar-graph on Worksheet One.

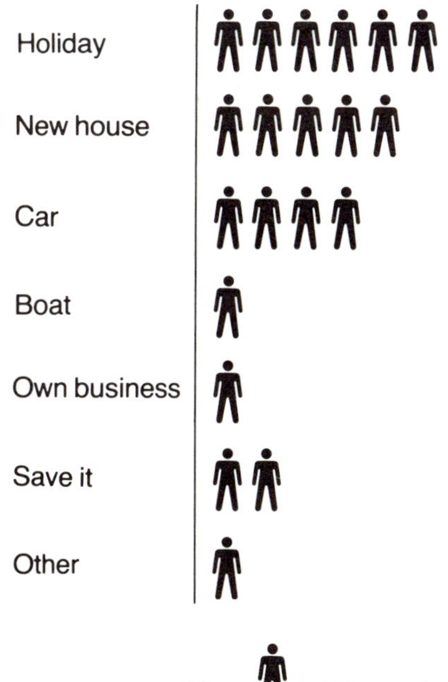

Figure 1

2 Look at the list of programmes on the television page shown on Worksheet Two. Put the programmes into categories. For example, soaps, news, documentaries, etc. If you cannot fit a programme into any category then put it in a column labelled 'others'.

3 Now you are going to conduct a survey to find out what sort of television programmes people like best. Write down some questions that will help you find out the facts about what people like to watch on television. You might be interested to find out if there are links between certain age groups, or between males and females.

4 Think carefully about the wording of your questions. Try some of them out on the people in your class to check the wording.

5 Now design a table to collect the results from your questionnaire. Decide roughly how many people you will be able to ask.

6 Go and collect your information.

7 When you have gathered your information draw pie-charts, bar-graphs and pictograms that can be easily understood.

8 Write a newspaper report to explain the results of your survey. You could use some of the diagrams you drew in step 7 to help make your points.

I've Won!

WORKSHEET ONE

Bar chart

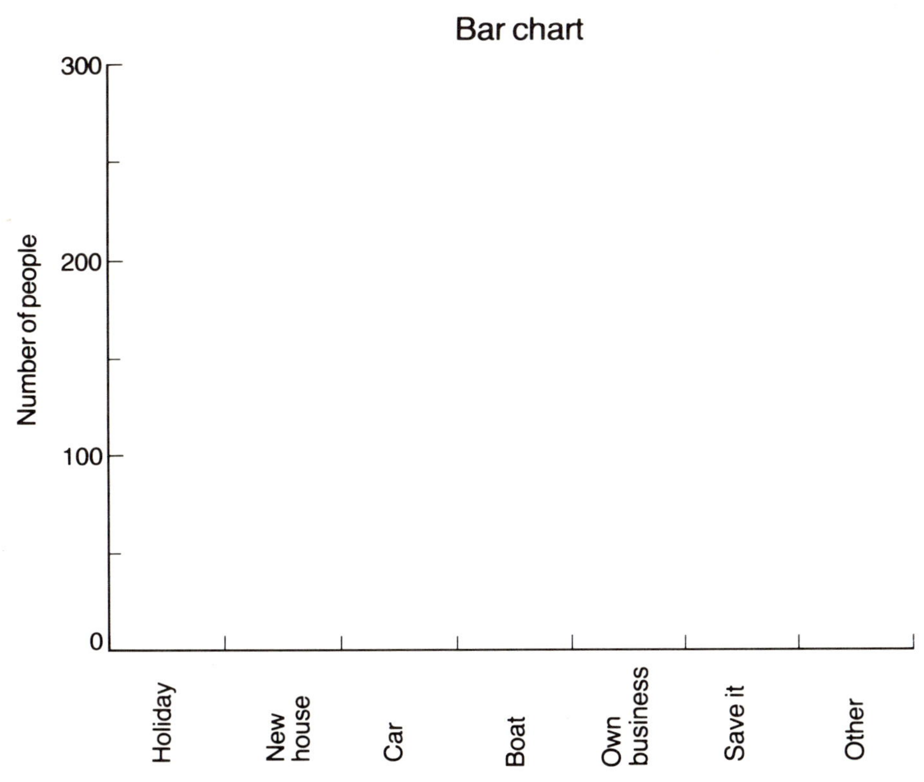

Pie chart

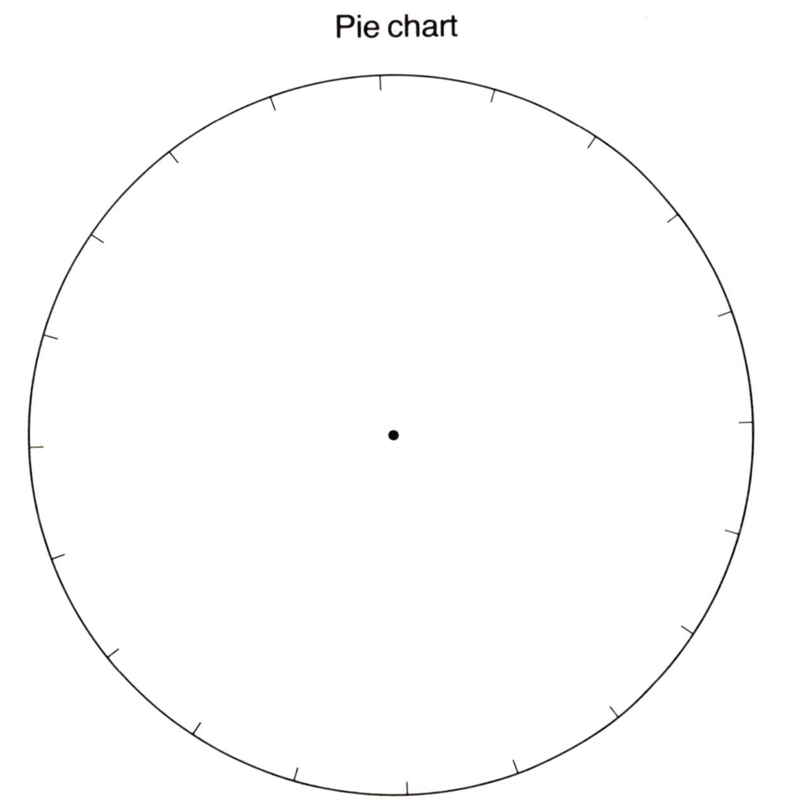

I've Won! WORKSHEET TWO

BBC-1

6 35 a.m. **LEON ERROL** in "Homework" (b/w). **6.55** Weather. **7** Breakfast Time with Frank Bough, Sally Magnusson, Jeremy Paxman and Pamela Armstrong. **8.35** The Pink Panther Show, rpt. **8.55**. Regional News and Weather. **9** News, Weather. **9.5** Children's BBC, and Charlie Brown. Rpt. **9.30** The Littlest Hobo, rpt. **10** News, Weather. **10.5** Neighbours rpt. **10.25** Childrens BBC and Play School, then The Adventures of Spot. **10.55** Five to Eleven. **11** News, Weather.

11 05 **GOLF—The Open.** The 116th Championship, from Muirfield, opening day. Including at **12** News and Weather; and at **12.55** Regional News and Weather.

1 00 **NEWS, WEATHER. 1.25** Neighbours. **1.50** Golf: The Open continued.

4 10 **THE KWICKY KOALA SHOW**—Repeat. **4.35** Silas, episode five. Rpt. (Ceefax). **5** John Craven's Newsround. **5.10** We Are the Champions; Heat 3, North of England. **5.35** Rolf Harris Cartoon Time, rpt.

6 00 **NEWS, WEATHER.**

6 35 **REGIONAL MAGAZINES.**

7 00 **TOP OF THE POPS** —Introduced by Peter Powell and Simon Bates.

7 30 **EASTENDERS**—Den is spotted driving off from the Queen Vic on a mysterious errand and the staff, preparing for the country and western night, hear nothing from him. The night succeeds, with a few surprises. (Ceefax).

8 00 **WHAT A CARRY ON!**—Still more funny fragments from the film series.

8 30 **DON'T WAIT UP**—Last of these repeats. Tom, reluctantly, has to give his father marching orders. (Ceefax).

9 00 **NEWS, WEATHER.**

9 30 **CRIMEWATCH UK**—Reconstructing the murder of a north London jeweller last month, only two days before he was to sell up and move away. Several people saw the shooting as he opened his shop. Also the murder of an 18-year-old girl in Newcastle-upon-Tyne. Someone could be sheltering the killer.

10 10 **BLACKADDER II**—Money. The date now is 1564, in England. Repeat.

10 40 **THE TROUBLE WITH SEX**—Someone, Somewhere Out There Cares. Agony aunts explain their work. Repeat.

11 10 **CRIMEWATCH UPDATE** —Developments from the earlier programme.

11 20 **THE ROCK GOSPEL SHOW**—With Sal Solo, Lavine Hudson and from America, Kathy Troccoli. Repeat. 12 **Weather**.

Wales—**5.35 p.m.-5.40** Wales Today. **6.35-7** Gardening Together. **12.00-12.05** News; Weather.

ITV Thames

6 15 a.m. **TV.am: GOOD MORNING BRITAIN,** 9.25 Thames News Headlines; followed by Schools.

12 00 **CREEPY CRAWLIES**—Repeat **12.10** Puddle Lane **12.30** The Sullivans. **1** News, **1.20** Thames News. **1.30** Falcon Crest: Aftershock, **2.25** Home Cookery Club: Saucy Apple Pudding, **2.30** Daytime, with Sarah Kennedy.

3 00 **FROCKS ON THE BOX**—High Street clothes. **3.25** Thames News Headlines. **3.30** Sons and Daughters. **4** Orm and Cheep. **4.10** Batfink, rpt.

4 20 **THE WIND IN THE WILLOWS**—A Producer's Lot. **4.45** The blunders, rpt. **4.50** Adventure of a Lifetime: Journey to Africa, with Matthew Kelly (Oracle). **5.15** Connections.

5 45 **NEWS.**

6 00 **THAMES NEWS**—With Andrew Gardner and John Andrew.

6 25 **HELP!**—With Viv Taylor Gee.

6 35 **CROSSROADS.**

7 00 **EMMERDALE FARM**—It's the local vicar, Donald Flinton, who has the inevitable problem today, following a visit from the bishop. Should he move on from Beckindale?

7 30 **NEVER THE TWAIN**—Words and Music Repeat. (Oracle).

8 00 **THIS WEEK**—Breaking Up the Family. A report on the "primary purpose" rule which restricts the entry to Britain of young men from India. Presented by Jonathan Dimbleby.

8 30 **ALL IN GOOD FAITH**—I Dreamt I Dwelt in Parish Halls. Trying to change parish hall bookings, vicar Richard Briers collides with the local martial arts club. (Oracle.)

9 00 **L.A. LAW**—According to the synopsis, there are at least six plot strands moving contrapuntally, including a woman broadcaster who is suing her employers for wrongful dismissal following a feature on her own breast surgery.

10 00 **NEWS AT TEN;** followed by **Thames News Headlines**

10 30 **ICE SKATING**—World figure Skating Championships. From Cincinnati.

11 30 **MELTDOWN**—Reggae night, with Big Audio Dynamite and Aswad.

12 30 **a.m. WORLD'S BEYOND** —Serenade for Dead Lovers. Nancy Travis and Jason Connery involved in a love affair of 40 years earlier.

12 55 **NIGHT THOUGHTS**—With the Rev. Barry Thorley.

I've Won!

ON YOUR OWN

1 Figure 1 shows three different graphs. Use the graphs to answer these questions.
(a) What was the class attendance for Wednesday?
(b) What percentage of her weekly budget does Sarah spend on food and rent?
(c) In which week did Kate have a score of 6?

2 Which sort of graph would you choose to show how the weight of a baby increases from birth?

3 If you wanted to know how many people in Great Britain watched a certain programme on television last night, how many people would it be reasonable for each of your survey team of four to ask?
(a) Up to 5 people;
(b) 5–100 people;
(c) 100–1000 people;
(d) over 10 000 people?

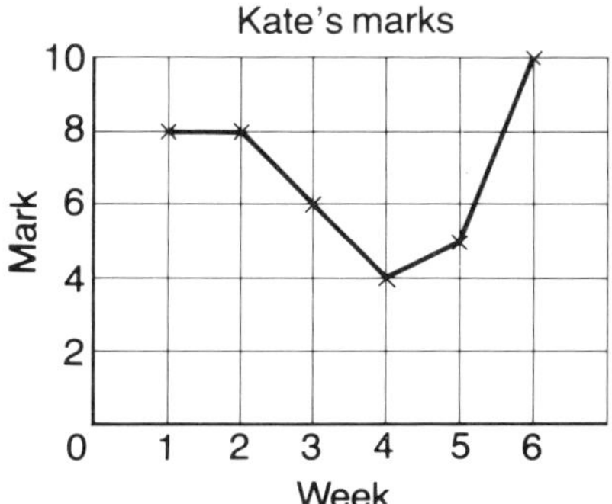

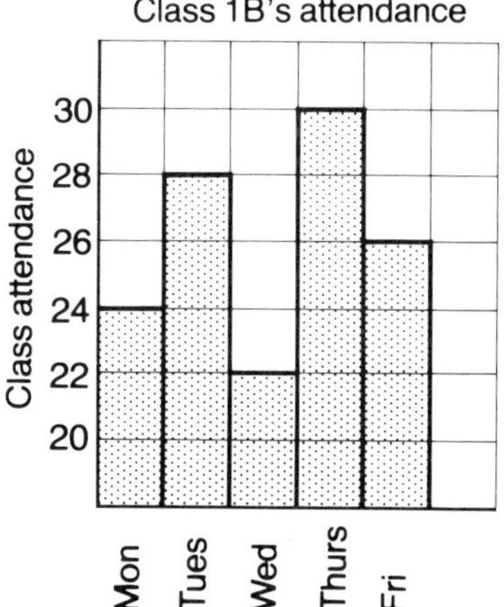

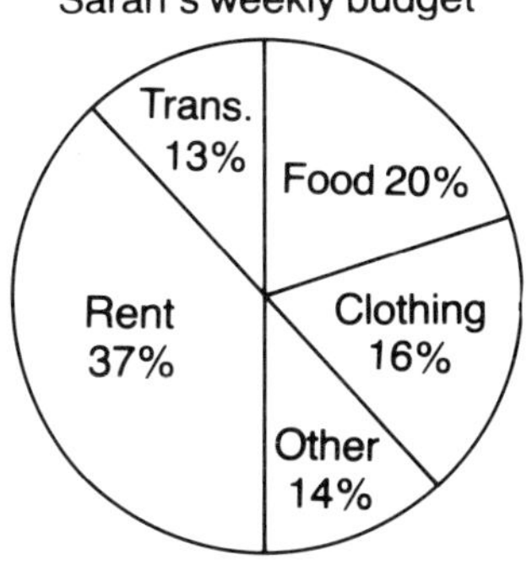

Figure 1

Countdown

Topic	Investigation.
Content	Practice with the number line and directed numbers. Searching for patterns and rules.
Assumed knowledge	Some experience of the number line being extended to negative numbers.
Integration	UM2 and SUM2: to precede *Integers: negative integers*. (See notes under Administration.)
Administration	Pairs or small groups. Revise the terms even, odd. Introduce or revise the terms positive and negative. Step 3 gives an approach that avoids the need to know how to subtract when negatives are involved, 'counting on' being used instead. More able pupils will probably realise that counting from -3 to 4 is the same as working out $4 - -3$, and may use a calculator to do this. Teachers should check that pupils are following the rule correctly, otherwise much wrong work may be generated. This is particularly important once negative numbers are involved. (If anyone can send us a clear, simple explanation for the 'final' set of numbers each time, we would be grateful!)

Answers and Marking Guide

Step	Notes	Max. Marks
1	(a) Check copy and insertion of the numbers 12, 18 and 12	3
	(b) See Figure T1	3
2	All examples should reduce to zero.	
	One example of all even:	
	Starting	1
	Completing	1
	One example of all odd:	
	Starting	1
	Completing	1
	One example of a mixture:	
	Starting	1
	Completing	1
	Writing about their discoveries	4

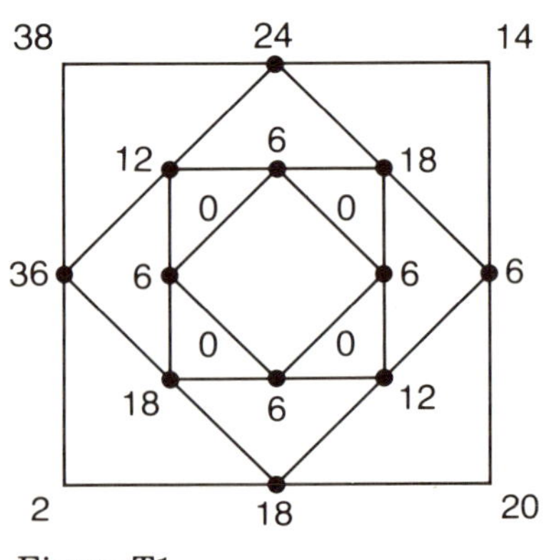

Figure T1

		Max. Marks
3	Correct completion of Figure 3 (see Figure T2)	2

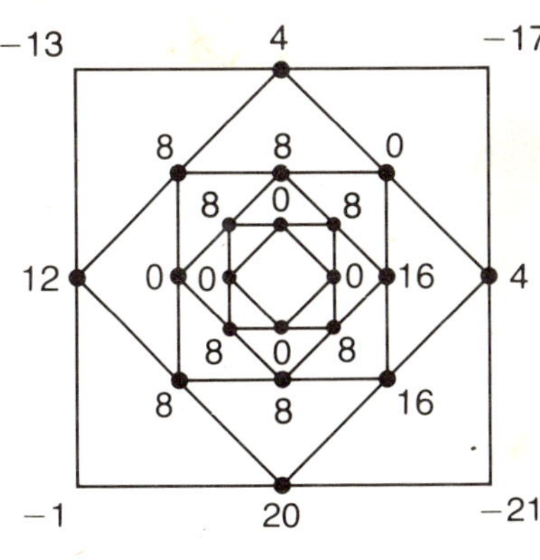

Figure T2

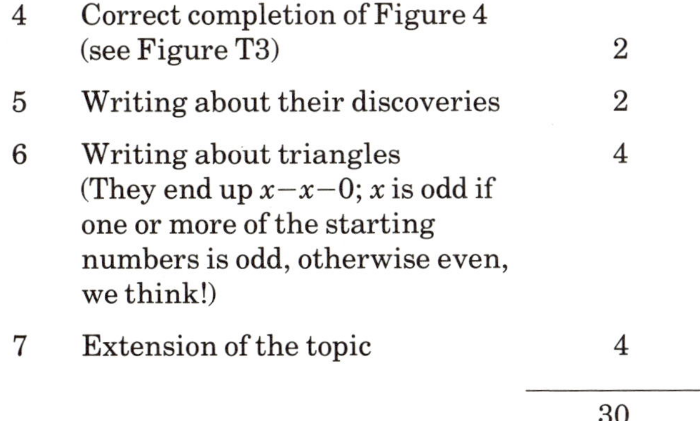

		Max. Marks
4	Correct completion of Figure 4 (see Figure T3)	2
5	Writing about their discoveries	2
6	Writing about triangles (They end up $x-x-0$; x is odd if one or more of the starting numbers is odd, otherwise even, we think!)	4
7	Extension of the topic	4
		30

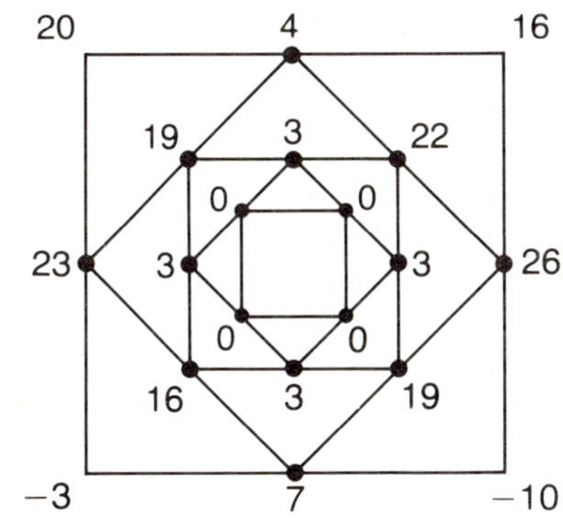

Figure T3

Suggested Grading Scheme

	A	B	C	D	E
Level 3	27+	24–26	20–23	18–19	16–17
Level 2	24+	20–23	17–19	15–16	13–14
Level 1	20+	17–19	14–16	12–13	10–11

On Your Own Answers

		Max. Marks
1	See Figure T4. One mark per correct square	4
2	Reduces to recurrent 1–1–0	4
3	The second square must be positive	2
		10

Figure T4

Countdown

1 Look at Figure 1. How did we work out the numbers 24, 6, 18 and 36 in Stage 2?

(a) Copy the diagram for Stage 3. Check that it is correct to write 18 between 6 and 24, then work out the other three numbers for the inside square and write them by the dots.

(b) Continue this pattern.

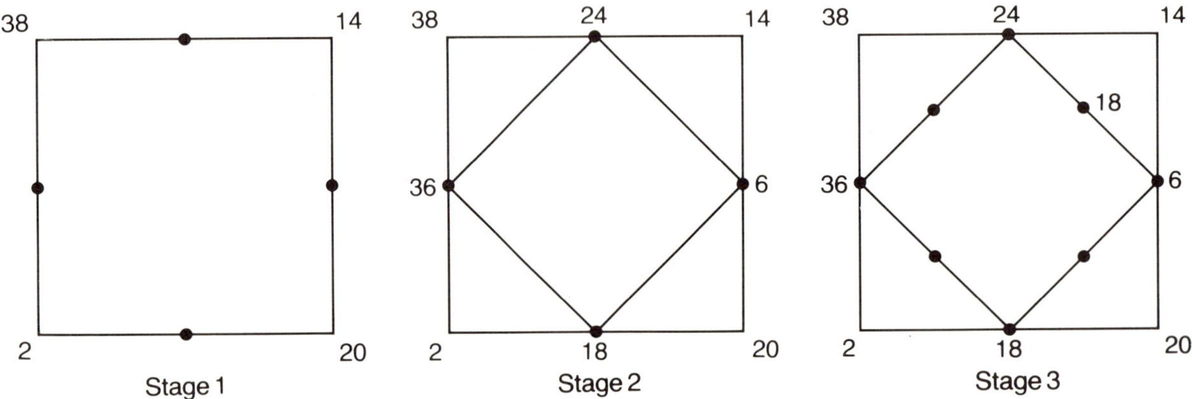

Figure 1

2 Now try the same idea with four different starting numbers. Try:

(a) all even;

(b) all odd;

(c) a mixture.

What else could you try? Write about your discoveries.

3 Figure 2 shows a number line from negative 12 to positive 12:

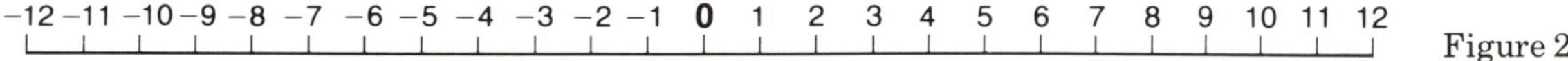

Figure 2

There are five divisions between 3 and 8.
There are five divisions between −3 and −8.
There are eleven divisions between −3 and 8.
There are eleven divisions between 3 and −8.

Look at Figure 3. A 4 has been written between −13 and −17 because there are four divisions between −13 and −17 on a number line. (If you have learnt how to subtract directed numbers, or have a calculator with +/− key on it, you may be able to find another way to get the answer 4 from −13 and −17.)

Copy Figure 3 and continue it as you did for Figure 1.

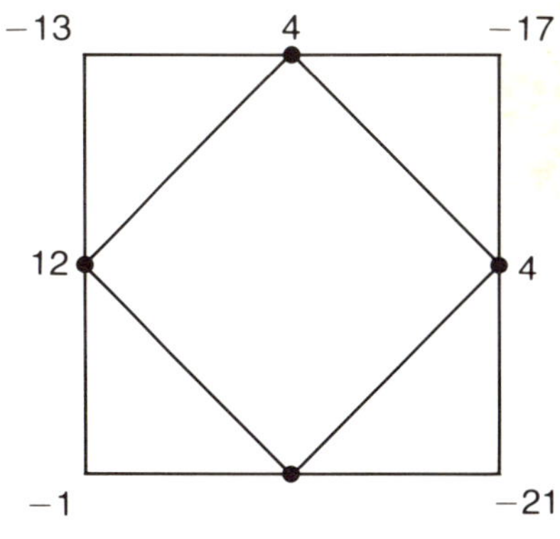

Figure 3

4 Copy and continue Figure 4.

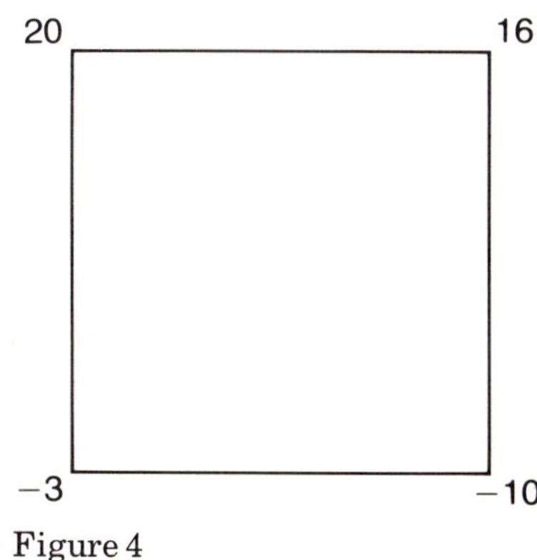

Figure 4

5 Write about the discoveries you made in steps 3 and 4.

6 Would the same idea work for other shapes? What about a triangle of three numbers? Investigate a triangle. Write about your discoveries.

7 What else could you do? You might try some other shapes, perhaps a hexagon, or a pentagon. Or you might do something else to develop the idea. It's up to you . . .

Countdown

ON YOUR OWN

1 Does the square in Figure 1 reduce to zero? Try it out.

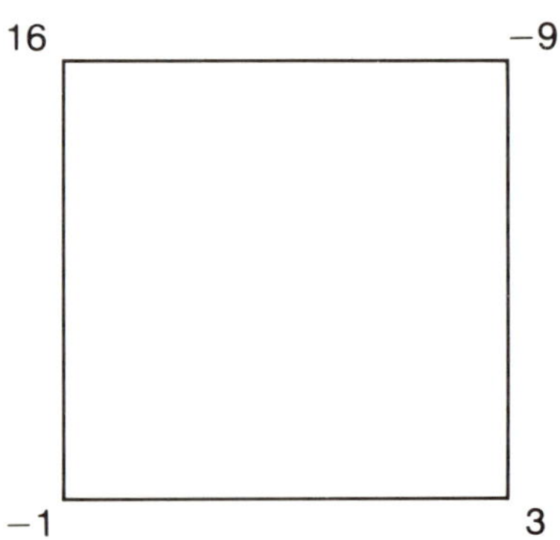

Figure 1

2 What about the triangle in Figure 2, does it reduce to zero? Is there any pattern at all?

3 What do you notice about the numbers on the second square you draw if you start off with all positive numbers, or all negative, or a mixture of both?

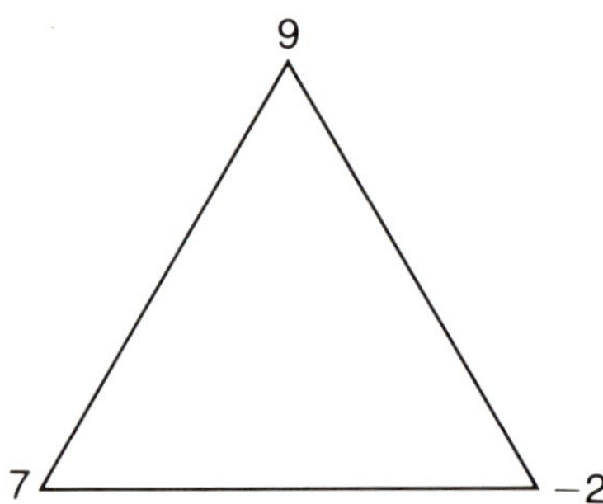

Figure 2

Fruity

Topic	Probability.
Content	Experimental probability. Organisation of data. Using probability to predict outcomes.
Assumed knowledge	Percentage symbol.
Integration	UM2 and SUM2: to precede *Probability: one event.*
Administration	Pairs. An envelope or pupil-made folder to keep the cut-out symbols in will be needed. Discussion of different results and conclusions is helpful.

Answers and Marking Guide

Step	Notes	Max. Marks
1–5	Collect data	3
6	Clear presentation of summary	3
7	Working out of profit or loss	3
8	(a) e.g. decrease the number of symbols used	2
	(b) e.g. increase the number of symbols used	2
9	Collection of results	1
10	Some idea of the implication of 50% profit	1
	Correct calculations	2
11	Ingenuity/practicality	3
		20

Suggested Grading Scheme

	A	B	C	D	E
Level 3	18+	15–17	12–14	9–11	7–8
Level 2	16+	14–15	11–13	8–10	5–7
Level 1	14+	12–13	9–11	6–8	3–5

On Your Own Answers

		Max. Marks
1	24 times	2
2	Not possible to score 1	2
3	Only one way to score 2 Only two ways to score 11 But six ways to score 7 And five ways to score 8	3
4	Loss is £1·70	3
		10

Fruity

1 The Worksheet shows some fruit-machine symbols (bell, melon, cherry, bar, banana, grapes, orange). Cut out the symbols and arrange them into three piles, those marked 1 in one pile, those marked 2 in another, and those marked 3 in a third. Shuffle each pile.

2 Prepare a result sheet headed:

Three the same	**Two the same**	**All different**

3 Look at the top three symbols. If they are all the same put a cross in the first column of your record sheet. If two are the same put a cross in the second column. If they are all different put a cross in the third column.

4 Take off the top symbol on each pile and repeat step 3. Keep doing this until you reach the bottom of the pile (ten crosses). Keep the three piles (1, 2 and 3) separate.

5 Shuffle the piles again and repeat steps 3 and 4 until you have 50 crosses.

6 Summarise your results clearly.

7 If every go on this fruit machine costs 10p, and the prizes are £1 for three the same and 25p for two the same, would you be in profit after your 50 goes? Show clear workings.

8 How could you change the reels on the fruit machine to (a) increase and (b) decrease the chances of winning?

9 Collect the results from 19 other pupils to give you 1000 results. (You only need to know their totals, as in step 6.) If there are fewer than 20 pupils in your class you will have to repeat steps 3 and 4 until you have enough to make 1000.

10 If it still costs 10p a go, what prizes could be offered, based on your results of 1000 goes, if the machine owner wants to make about 50% profit?

11 Design your own game or machine which could be used at a school fête. It should be based on chance and you should work out prizes and what profit you would like to make.

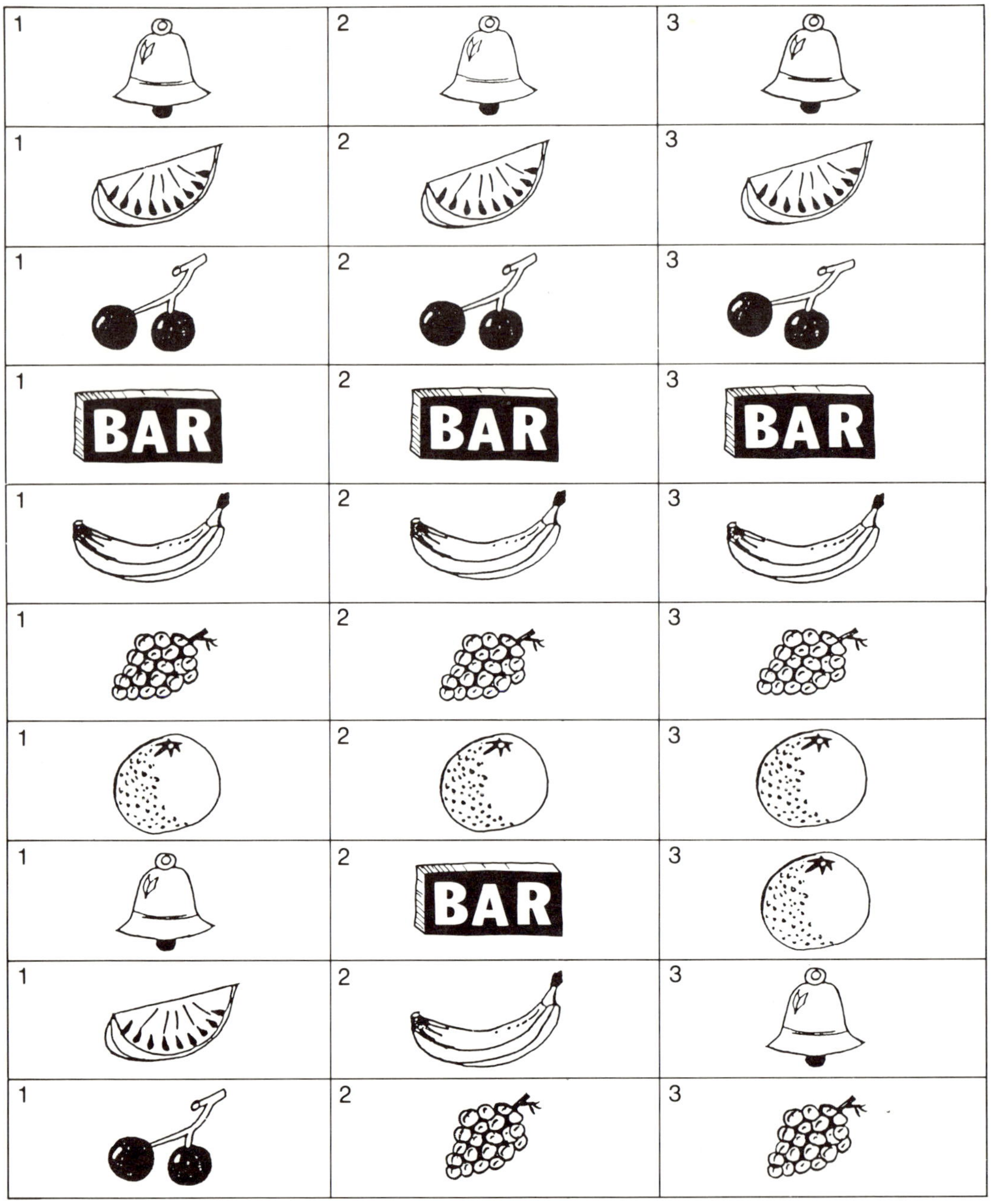

Fruity

WORKSHEET

1	2	3
1	2	3
1	2	3
1 BAR	2 BAR	3 BAR
1	2	3
1	2	3
1	2	3
1	2 BAR	3
1	2	3
1	2	3

ON YOUR OWN

Look at the information in the table below which shows you the results obtained when rolling two dice 100 times.

Total score	Number of times scored
2	4
3	10
4	10
5	14
6	13
7	16
8	16
9	8
10	4
11	3
12	2

1 How many times did I score four or less?

2 Why didn't I put a score of 1 on my table?

3 Why do I score more 7's and 8's than 2's or 11's?

4 If I was given 5p for every time I threw a 10, 11, 12, 2, 3 or 4 and paid 5p for every 5, 6, 7, 8, or 9 would I make a loss or a profit? How much?

Speed Limits

Topic	Everyday application.
Content	Use of conversion graphs. Dials and scales. Rounding.
Assumed knowledge	Measurement with ruler. Use of protractor to divide circle into equal sections.
Integration	UM2: to precede *Conversion graphs.* SUM 2: to precede *Using your calculator: keeping slim.*
Administration	Individual or small groups. 360° protractors will be required.

Answers and Marking Guide

Step	Notes	Max. Marks
1	(a) 32 km (1 mark for wrong answers within 5 units either way)	2
	(b) 10 miles	2
2	Presentation	2
	(a) 15 (b) 50 (c) 80 (d) 110 (allow 115) Answers correct, 2 marks each. Award 1 mark if not rounded	8
3	Presentation	2
	Correct division into 16 parts	2
	Accurate measurement	2
	Correct numbering	2
	Note: A tracing or OHP transparency is useful to check this (see Figure T1)	

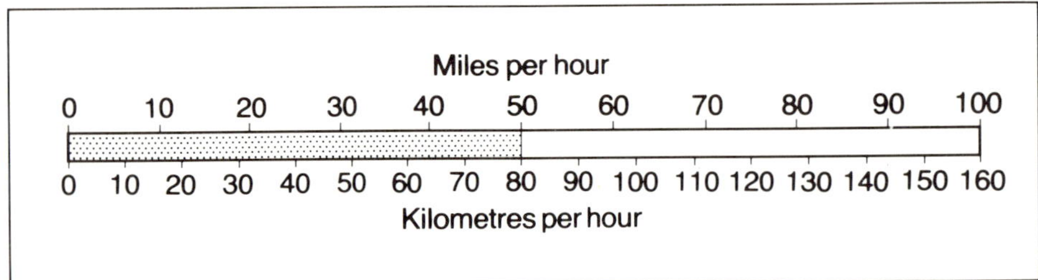

Figure T1

		Max. Marks
4	Presentation	2
	Correctly dividing both scales	2+2
	Correctly numbering both scales	2+2
	Note: A tracing or OHP transparency is useful to check this (see Figure T2)	
5	Inventiveness	4
	Practicality/usefulness	4
		40÷2
	Total: Divide by 2 and round up.	20

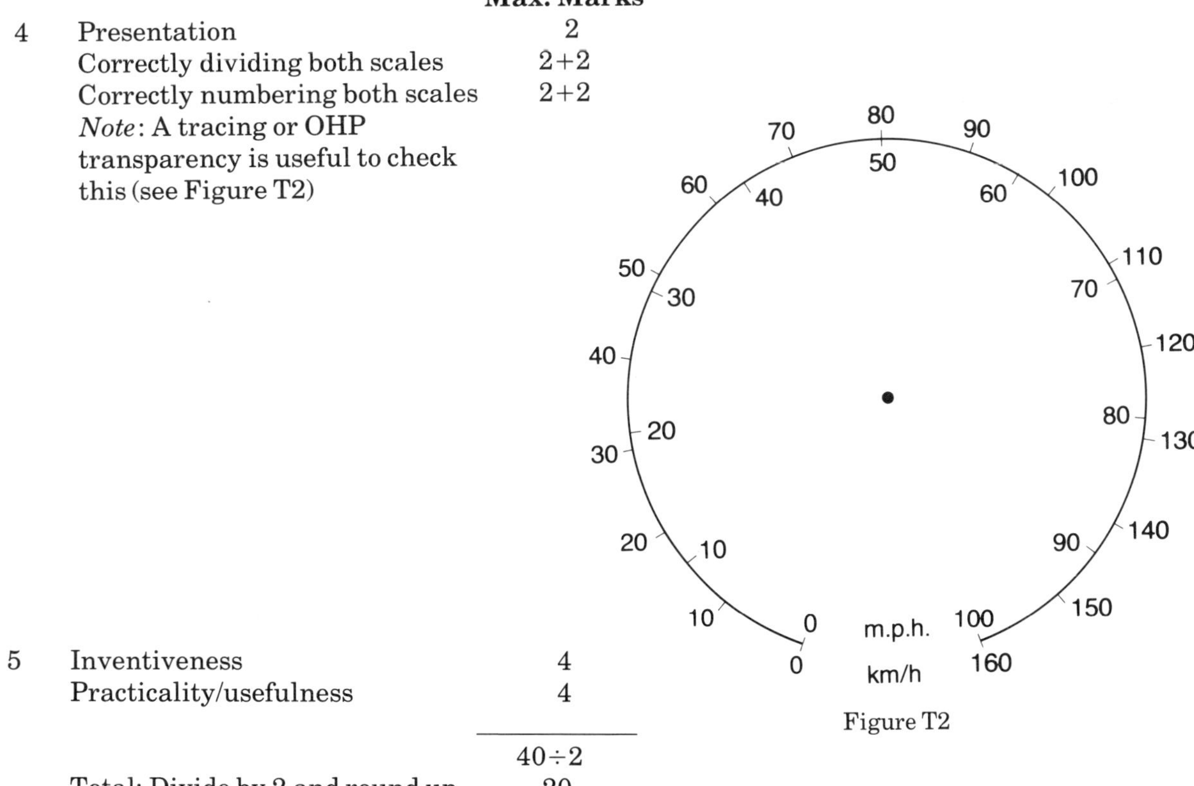

Figure T2

Suggested Grading Scheme

	A	B	C	D	E
Level 3	19+	16–18	13–15	11–12	9–10
Level 2	16+	13–15	10–12	8–9	6–7
Level 1	14+	11–13	8–10	6–7	4–5

On Your Own Answers

		Max. Marks
1	(a) 22 pounds (1 mark for 20 to 24 pounds)	2
	(b) 30 kg (1 mark for 29 to 31 kg)	2
	(c) 45 kg (1 mark for 40 to 50 kg)	2
2	Kilograms should be marked every 11 divisions See Figure T3	4
		10

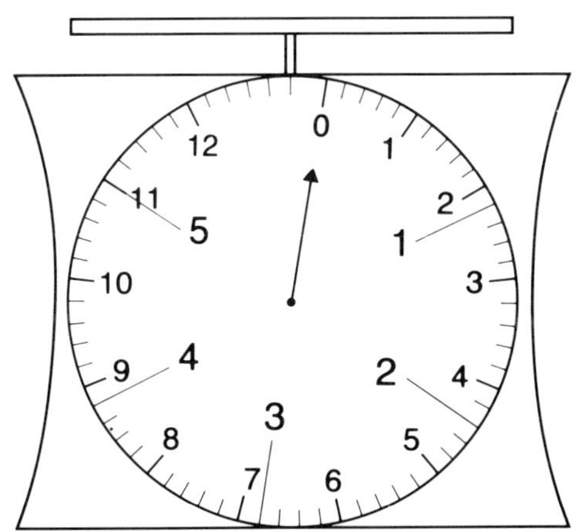

Figure T3

Speed Limits

1 Figure 1 is a conversion graph that converts miles into kilometres, or kilometres into miles. From the graph convert:
(a) 20 miles into kilometres;
(b) 16 kilometres into miles.

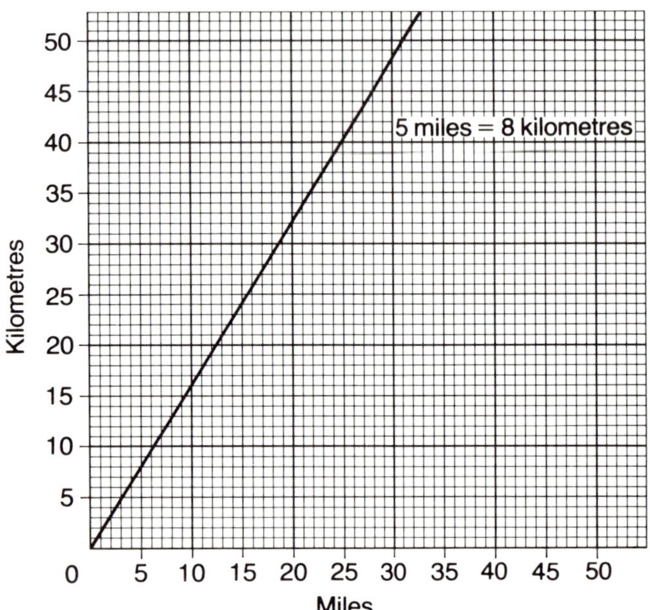

Figure 1

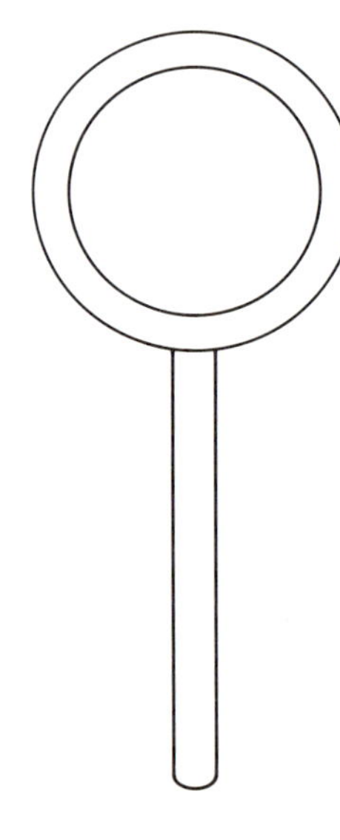

Figure 2

2 Copy the road sign in Figure 2 four times. Convert these speeds in miles per hour to the nearest 5 kilometres per hour and write the kilometre values in your signs:
(a) 10 m.p.h;
(b) 30 m.p.h;
(c) 50 m.p.h;
(d) 70 m.p.h.

3 On the Worksheet is a straight line speedometer. The speedometer reading is shown by a solid band. At present it shows 50 miles per hour. Complete the speedometer by marking and numbering the kilometre side of the scale so that a motorist could use it in France. Use multiples of ten, 0, 10, 20, 30 . . .

4 Look at the circular speedometer on the Worksheet. Fill in the miles and kilometres per hour so that it can be read accurately. You should only mark whole numbers in multiples of 5 or 10 for both the miles and kilometres.

5 Figure 3 is a conversion graph between gallons and litres. Design something interesting that could be given to motorists to help them convert the amounts of petrol they are buying from litres to gallons and from gallons to litres.

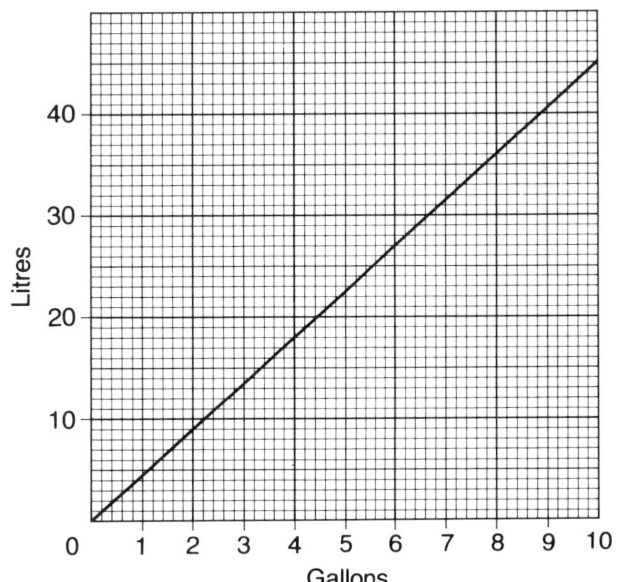

Figure 3

Speed Limits WORKSHEET

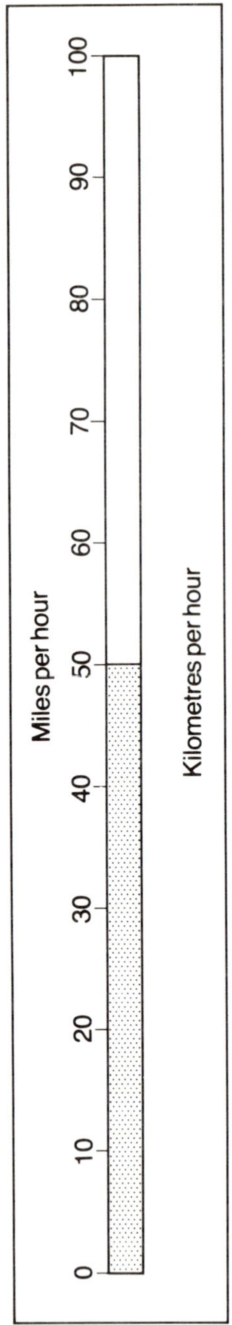

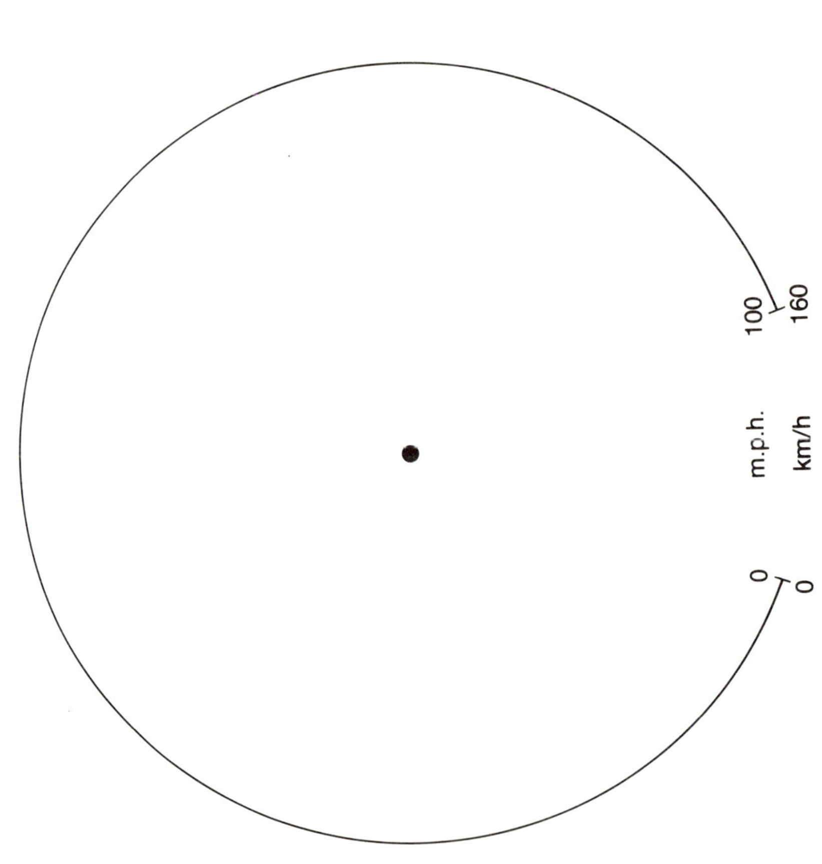

Speed Limits ON YOUR OWN

1 Look at Figure 1. It shows a conversion graph between weights in pounds and weights in kilograms. Answer the following questions:
(a) How many pounds are the same as 10 kilograms?
(b) How many kilograms are the same as 66 pounds?
(c) Estimate how many kilograms would be the same as 99 pounds.

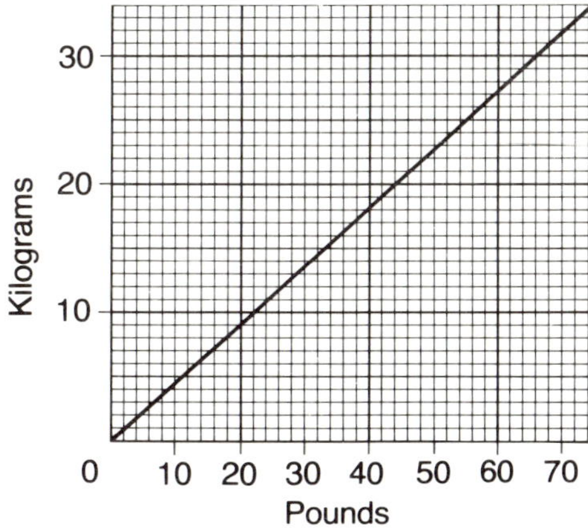

Figure 1

2 Look at the weighing scales in Figure 2. The pounds have been marked in. Mark the scale so that it could be used to weigh whole kilograms from 1 to 5.

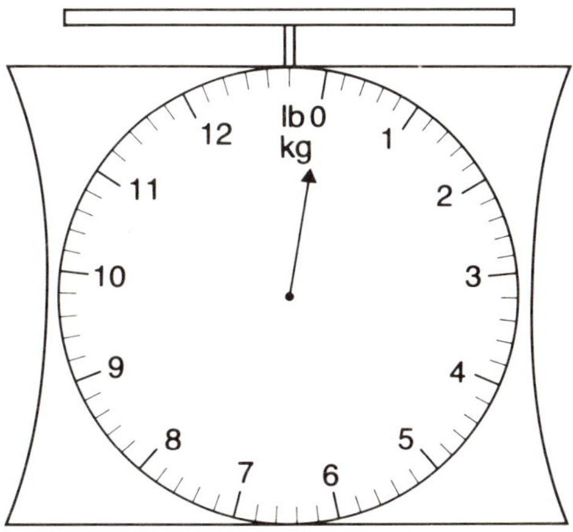

Figure 2

Nets

Topic	Practical geometry.
Content	Making models from nets. Drawing and measuring nets. Naming solids. Maximising. Investigating permutations of positioning of solids. Finding rules.
Assumed knowledge	Names of polygons. Use of ruler and protractor.
Integration	UM2/SUM2: part of *Project: Constructing prisms.*
Administration	Small groups. Thin card, glue and scissors will be needed (A4 size card is adequate). Encourage pupils to test answers by making the shapes for themselves, but they should write down their answers and findings, or else you will be left with the problem of marking hundreds of cubes and pyramids! Nets that have been cut out can be stored flat in the envelope made in question 1. Some pupils will need guidance in scoring (with a ball-point), accurate folding, and the tab gluing sequence. Be patient!

Answers and Marking Guide

Step	Notes	Max. Marks
1	Envelope construction	2
2	(a) No marks for this, unless you mark as they are made	
	(b) Nets 4, 5 and 7 make cubes	3
3	It will not make a pyramid	1
	Clear reasons	2
4	Correct net	2
5	Either 18 cm by 16 cm or 8 cm by 23 cm	2
	Giving both answers: bonus	1
6	(a) Correct net	2
	(b) Correct net	2
7	There are 24 views, see Figure T1. Up to 10 answers: 2 marks Up to 20 answers: 3 marks All 24 answers	4
8	(a) (i) 5 nets	2
	(ii) 2 nets	2

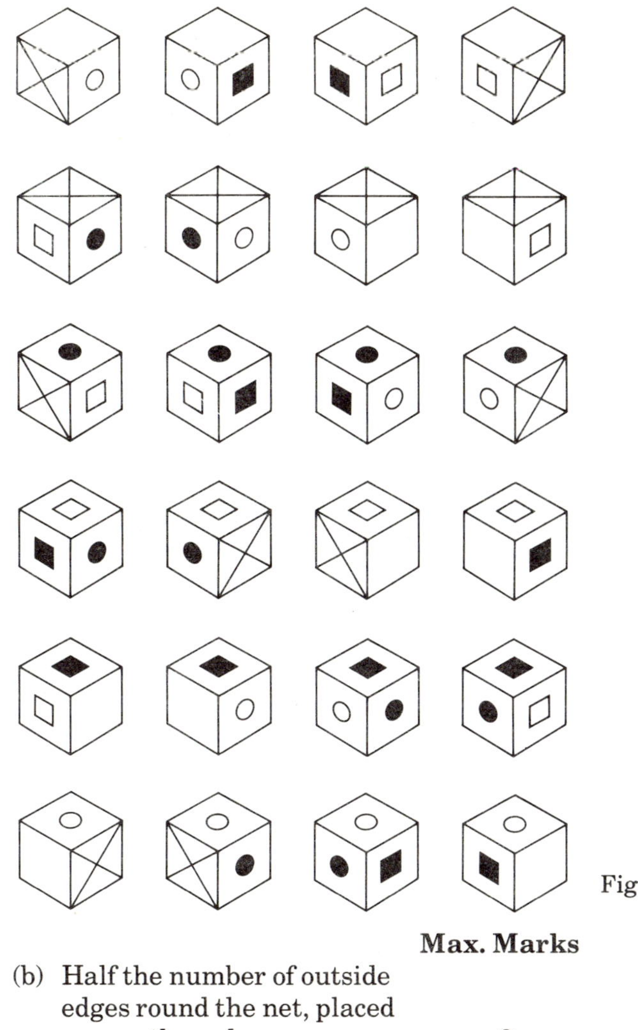

Figure T1

		Max. Marks
	(b) Half the number of outside edges round the net, placed every other edge	2
9	Efforts and interest shown	3
		30

Suggested Grading Scheme

	A	B	C	D	E
Level 3	25+	22–24	19–21	16–18	14–15
Level 2	20+	18–19	15–17	13–14	11–12
Level 1	16+	14–15	12–13	10–11	8–9

On Your Own Answers

		Max. Marks
1	Correct net	3
	Accurate drawing	2
2	(a) 5	1
	(b) Correct net	3
	Accurate drawing	1
		10

Nets

1 Figure 1 is the net of an envelope. Use a piece of A4 paper to make it. Cut off the shaded parts and fold and stick the side tabs.

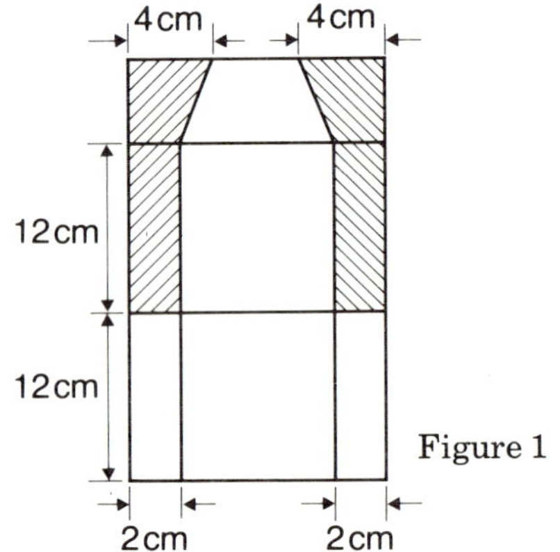

Figure 1

2 A cube has six faces and each one is the same size square. (See Figure 2.)
(a) Number 1 on Worksheet One is the net of a cube. Cut it out and carefully bend the tabs, then fold and stick until it becomes a cube.
(b) Look at the other nets on the sheet. Try to work out which ones will make a cube. Test your answers by cutting them out and folding them.

3 Figure 3 shows a pyramid. Look at the net in Figure 4. Will it make a pyramid? Give a reason for your answer.

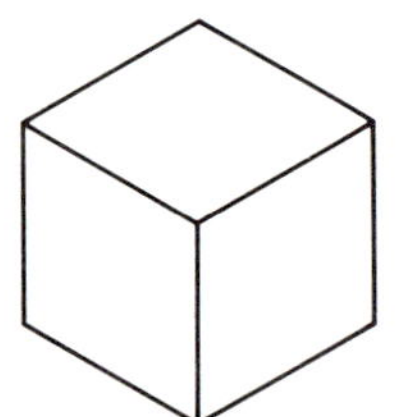
Figure 2

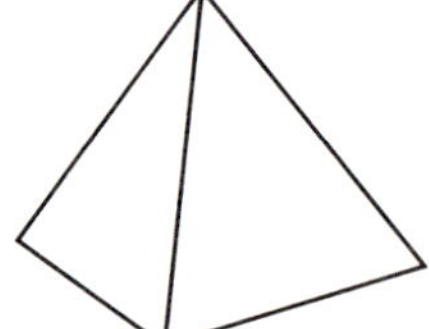
Figure 3

4 The Egyptian pyramids have a square base. Draw a net for a square-based pyramid.

5 A manufacturer wants to make chocolate boxes in the shape of a tetrahedron (a triangular based pyramid). He has worked out that the sides of the triangle all have to be 9 cm. Tabs are to be 1 cm wide. By drawing a net of this tetrahedron work out the smallest size rectangle of card he would need for one box.

6 A pyramid can have a base with as many sides as you like. Draw the net for a pyramid with:
(a) a regular five-sided base;
(b) a regular six-sided base.
Cut out and fold up your nets to check them.

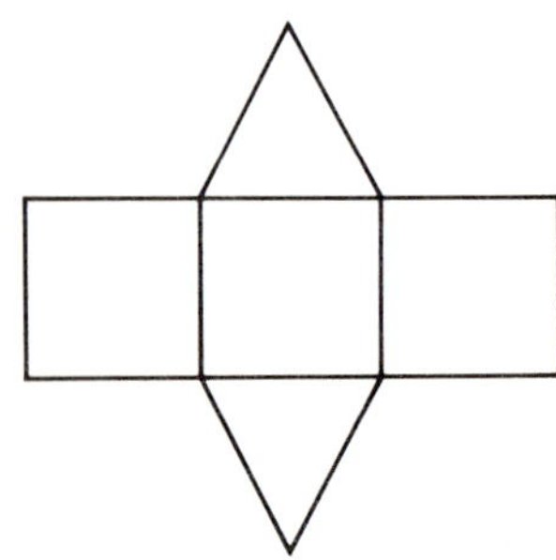
Figure 4

7 Figure 5 is the net of a cube. Each face is different.

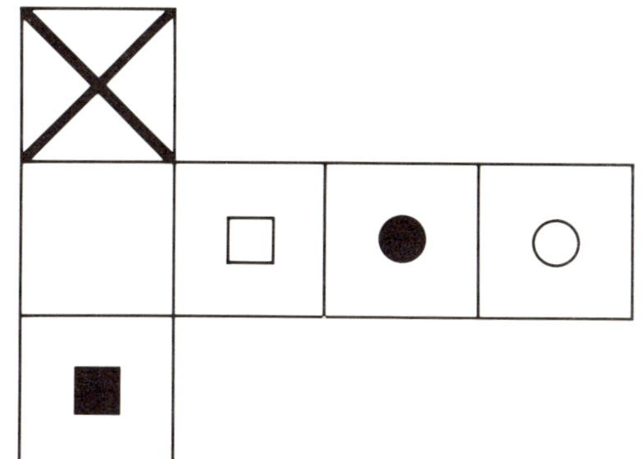

Figure 5

Figure 6 shows one way to place the cube on the desk. There are more than 20 different ways of placing the cube. Use Worksheet Two to draw all the different ways you can find. How many different ways are there? Why?

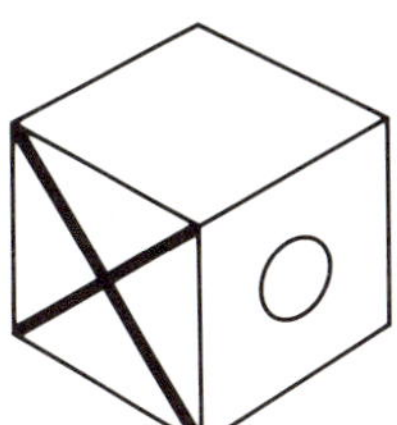

Figure 6

8 Nets and tabs

(a) How many different nets are there for (i) a cube; (ii) a tetrahedron?

(b) Can you find a rule for how many tabs any net requires and where they should be placed?

9 You might like to make models of the five platonic regular solids. They are:

regular tetrahedron – 4 equilateral triangles;
cube – 6 squares;
octahedron – 8 equilateral triangles;
dodecahedron – 12 regular pentagons;
icosahedron – 20 equilateral triangles.

Nets

WORKSHEET ONE

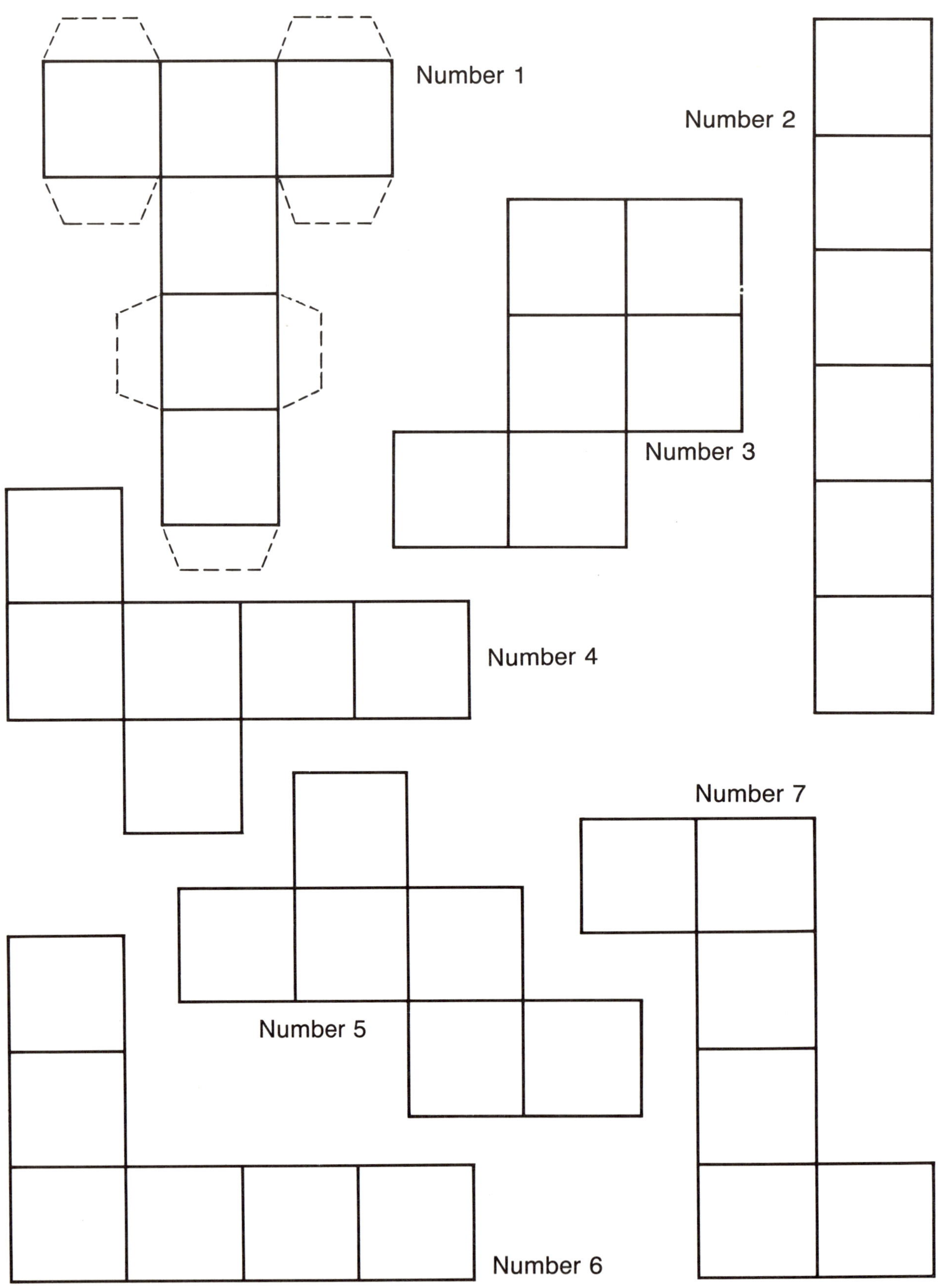

Nets

WORKSHEET TWO

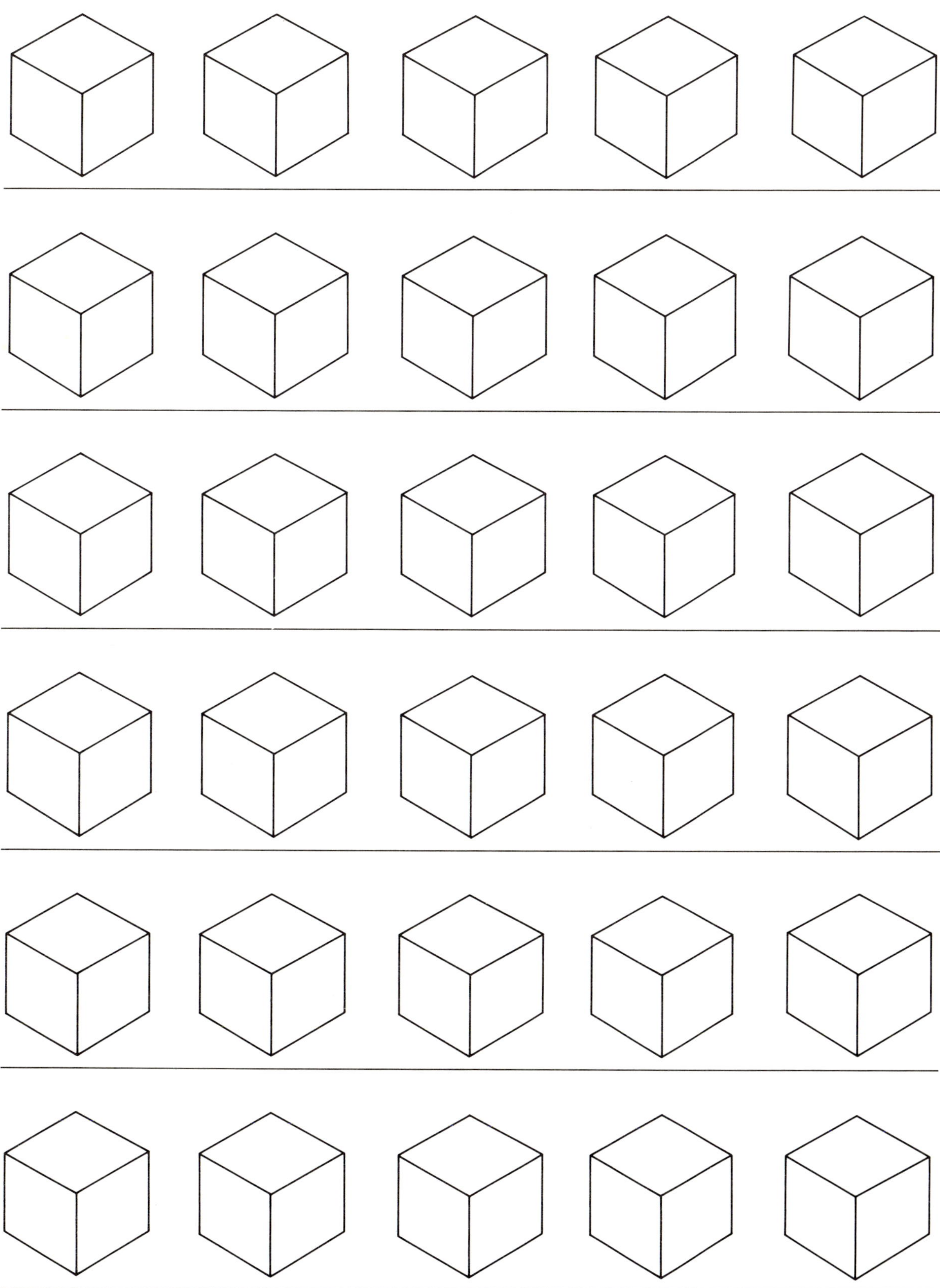

Nets

ON YOUR OWN

1 Figure 1 is the net of a cuboid. Can you draw a different net that will make the same cuboid?

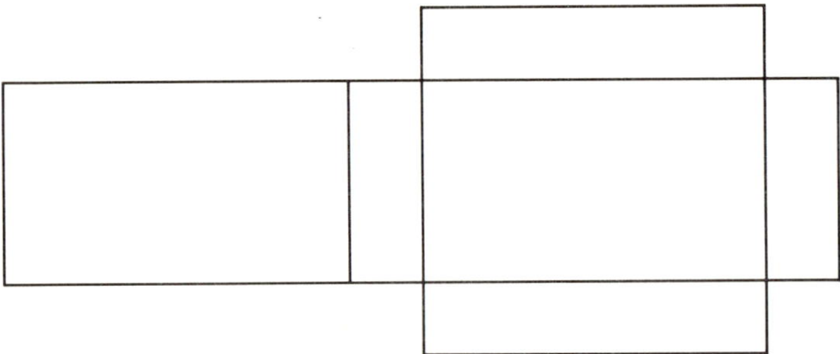

Figure 1

2 The solid in Figure 2 is called a triangular prism.
(a) How many faces does it have?
(b) Draw a net that will make a triangular prism.

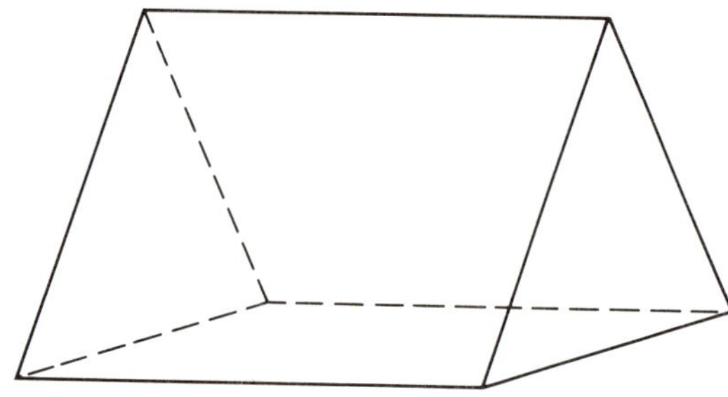

Figure 2

Hold the Line

Topic	Problem-solving.
Content	Measuring. Interpreting scales.
Assumed knowledge	Measurement.
Integration	UM2 and SUM2: after *Metric system* problems.
Administration	Group discussion or individual work. Some of the more able pupils may wish to do exact calculations of the length of cable and some may note that a telegraph pole is taller than a bungalow and use Pythagoras' theorem, deciding for themselves the average height of a telegraph pole and a bungalow.

Answers and Marking Guide

Step	Notes	Max. Marks
1	Any sensible position. At least three positions should be considered along with the measurements. One reason for choice is obviously the least amount of cable. Other reasons for choice will be not blocking roads, windows, pathways, etc.	10
2	Again, sensible positions using the least amounts of cable	5
		15

Suggested Grading Scheme

	A	B	C	D	E
Level 3	14+	12–13	10–11	8–9	6–7
Level 2	12+	10–11	8–9	6–7	4–5
Level 1	10+	8–9	6–7	4–5	2–3

On Your Own Answers

		Max. Marks
1	12·5 cm scaled; 25 m true	3
2	6·5 cm scaled; 13 m true Length of pipe saved = 15 metres	3
		6

Hold the Line

Sunningdale Park is an estate of small bungalows for old-age pensioners. Each of the bungalows is to have a telephone and the telephone company must decide where to put the new telegraph poles. The Worksheet shows a plan of Sunningdale Park.

The telephone company must bear in mind that each cable needs to be connected to one of the corners of the roof of each bungalow. There are to be no party lines and each bungalow must be connected separately to the telegraph pole.

1 If they are going to use just one telegraph pole where should it be sited? Consider several positions and give reasons for your choice.

2 If the company decides to put up two telegraph poles where should they site them?

Hold the Line

WORKSHEET

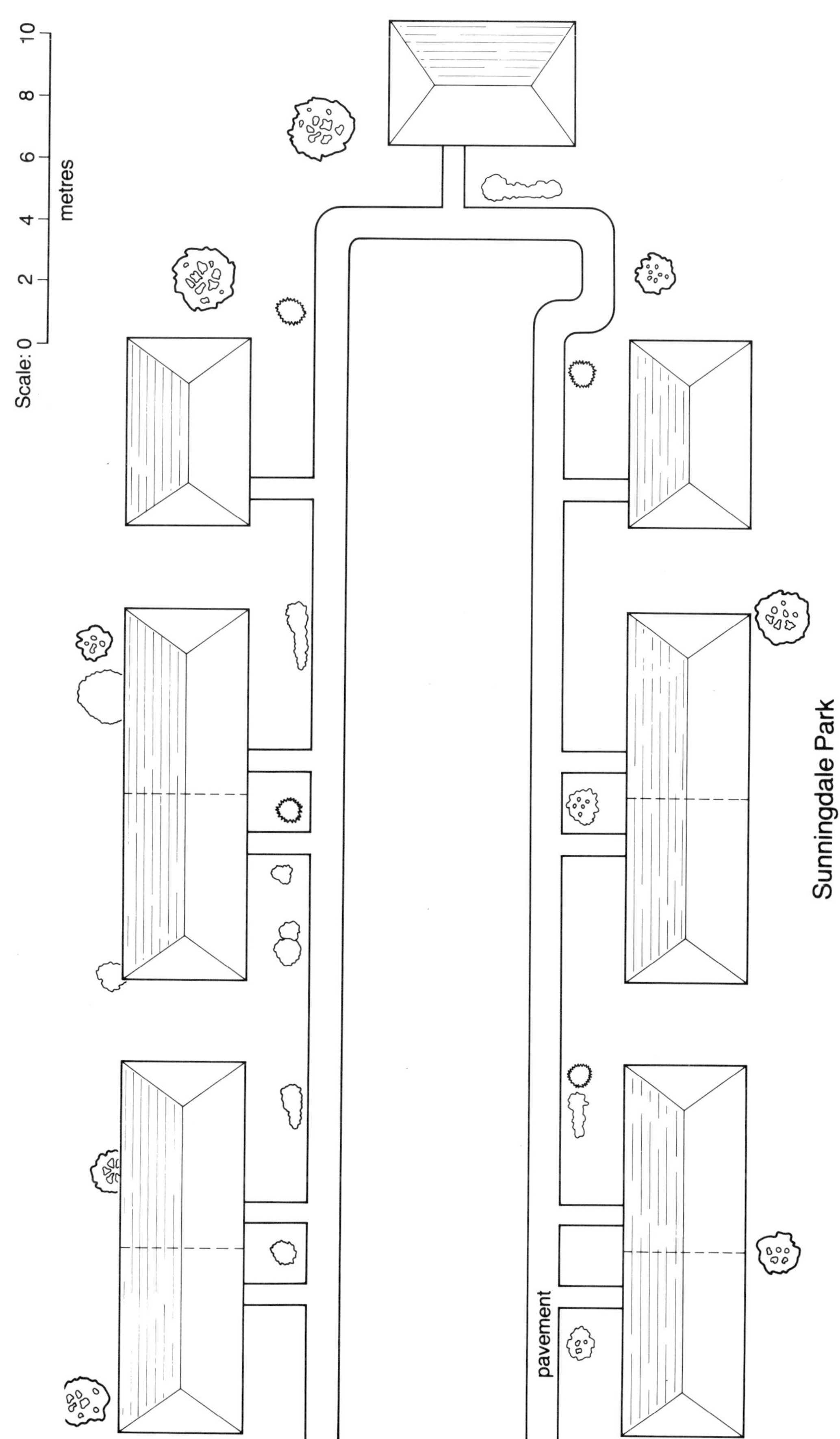

Hold the Line

ON YOUR OWN

1 On the plan in Figure 1 what length of pipe will be needed to connect an underground water supply directly to the house from the mains supply, if you have to follow the route of the road and driveway?

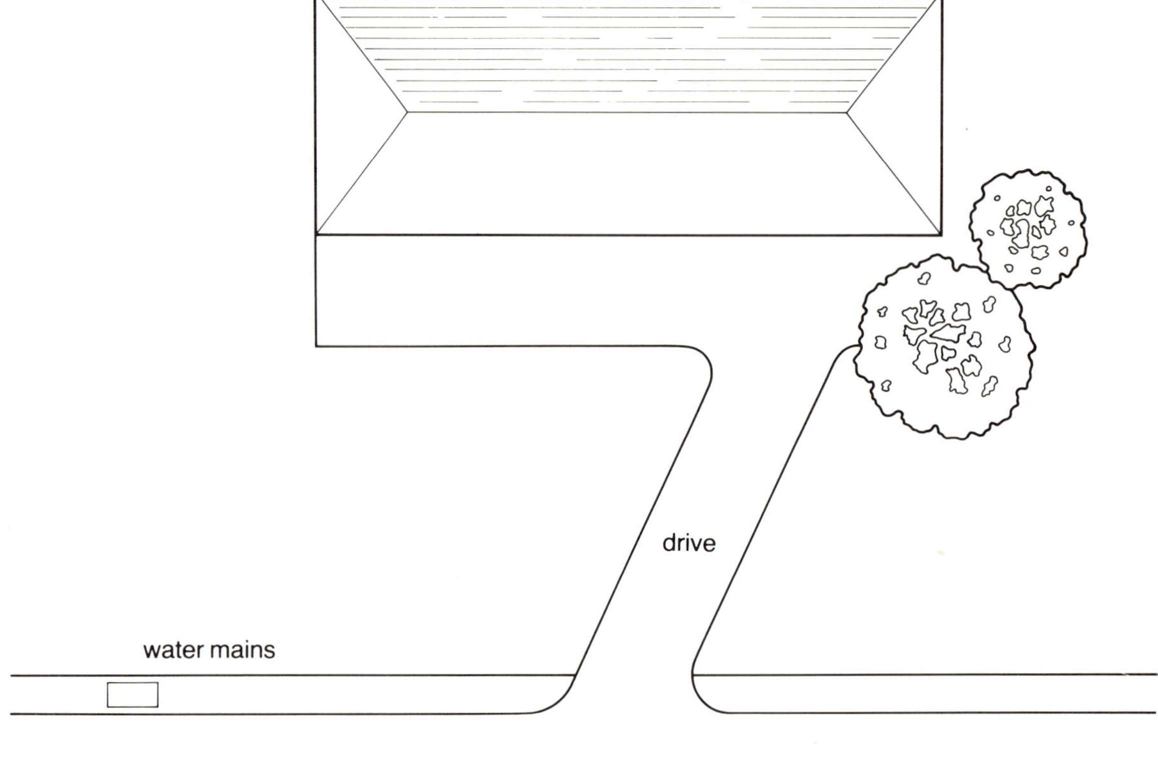

Figure 1

2 If you could connect the house straight to the mains without following any particular route how much pipe could you save?

A Day Out

Topic	Everyday application.
Content	Planning a journey. Drawing Time/distance graphs. Reading information from tables.
Assumed knowledge	Time/distance graphs.
Integration	UM3: to follow *Time/distance graphs.* SUM3: to follow *Travel graphs.*
Administration	A group discussion is necessary first with careful planning of the journey. Pupils will have to assume some information such as the length of time of the ferries on the Portsmouth route. Step 2 encourages pupils to find their own information. For the purpose of this application British Summer Time has been ignored. It could be included if wished.

Answers and Marking Guide

Step	Notes	Max. Marks
1	Portsmouth/Le Havre, see Figure T1. Look for a graph that is clear and assess accuracy. Time (at least 30 minutes) should be allowed to board the ferry. Time may be allowed to pass through customs on return.	5

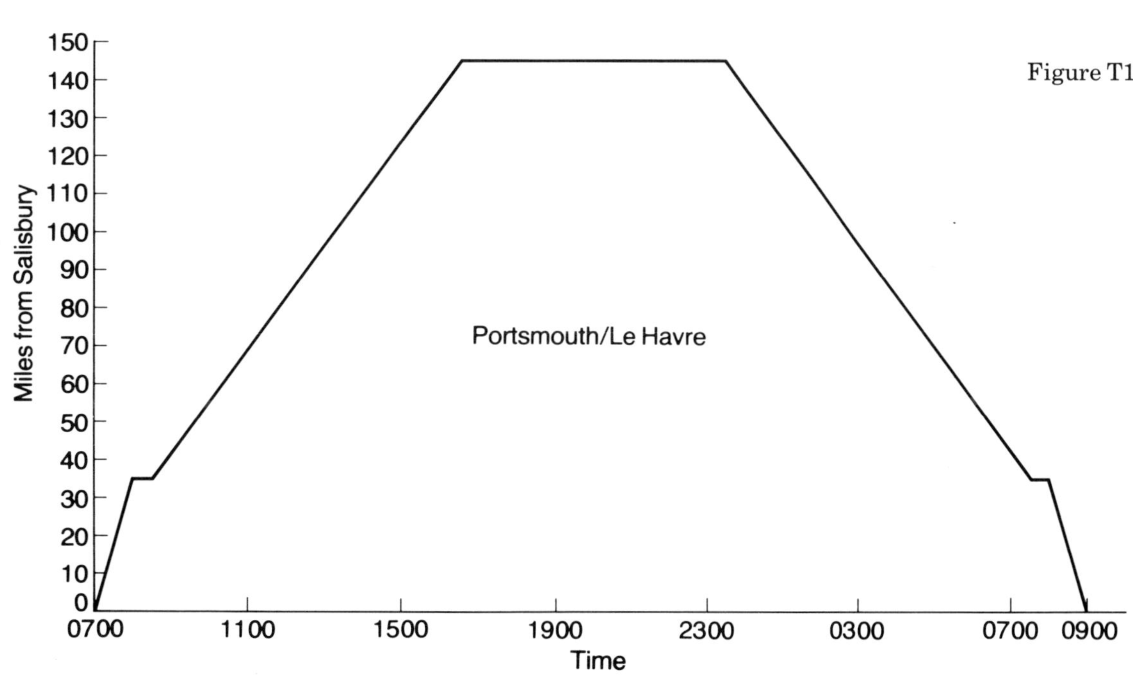

Figure T1

	Max. Marks
Figure T2 is an example for the Newhaven/Dieppe journey, although it will depend on the times chosen	5
Comparisons Any along these lines: (a) Further to Newhaven so car journey is more expensive. (b) Length of time ashore. (c) Is the Dieppe route cheaper (children's fares, duty-free goods, 'from £9·00')?	5

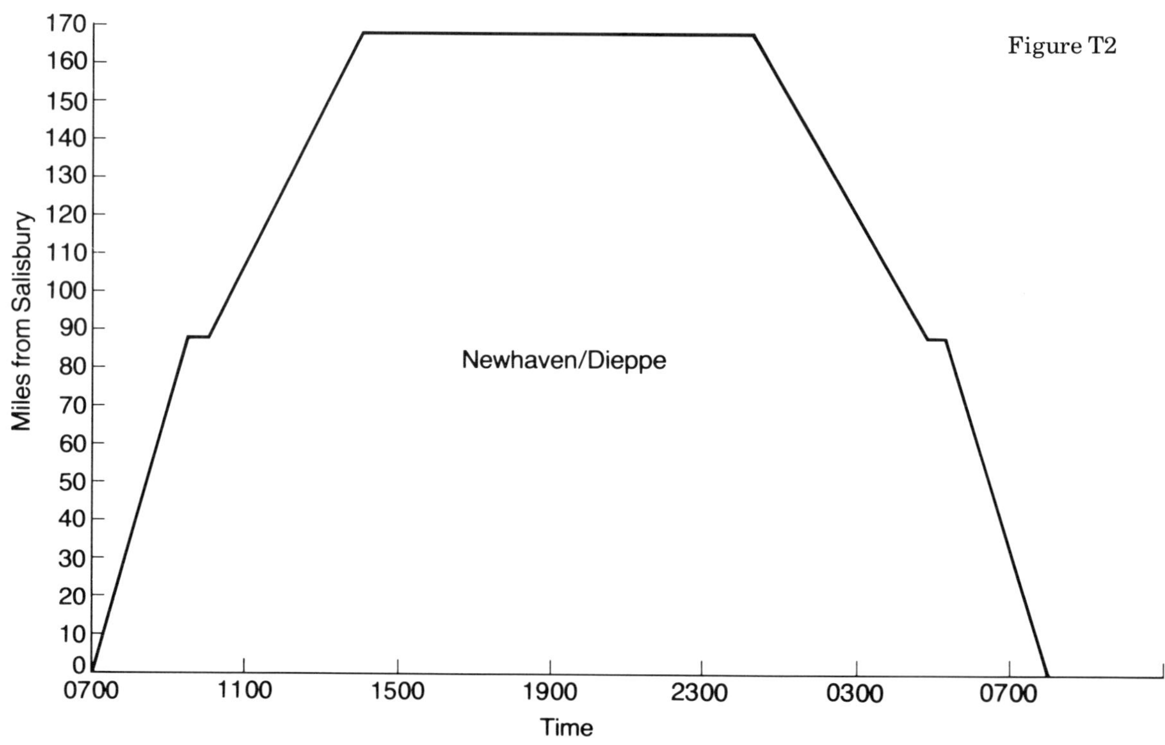

Figure T2

2	Planning own journey. Check for accuracy. Presentation of all the information and a time/distance graph	5
		20

Suggested Grading Scheme

	A	B	C	D	E
Level 3	17+	15–16	12–14	10–11	8–9
Level 2	15+	12–14	10–11	8–9	6–7
Level 1	12+	10–11	8–9	6–7	4–5

On Your Own Answers

		Max. Marks
1	Two journeys	2
2	1 hour 45 minutes	2
3	9:15 a.m. (0915)	2
4	The buses pass on the road.	2
5	It stops for 15 minutes.	2
		10

A Day Out

1 From the information on the Worksheet you are going to plan a day trip to France. You will be starting from Salisbury by car and you have a choice of two ferry routes, either Portsmouth to Le Havre or Newhaven to Dieppe. Your task is to plan journeys for both options, including in your planning a time/distance graph. Then you should compare your two journeys and comment on them. See Figures 1 and 2.

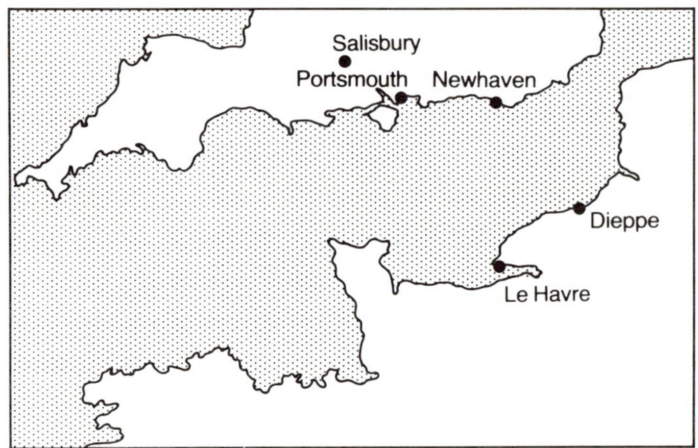

Figure 1

You should include the journey back to Salisbury.

Plan the journeys as they should happen if all goes smoothly.

Assume an average road speed of 35 miles per hour.

2 Plan any journey from your home. You may use any method of transport but it should be completed in 24 hours. Again you should include a time/distance graph as part of your planning.

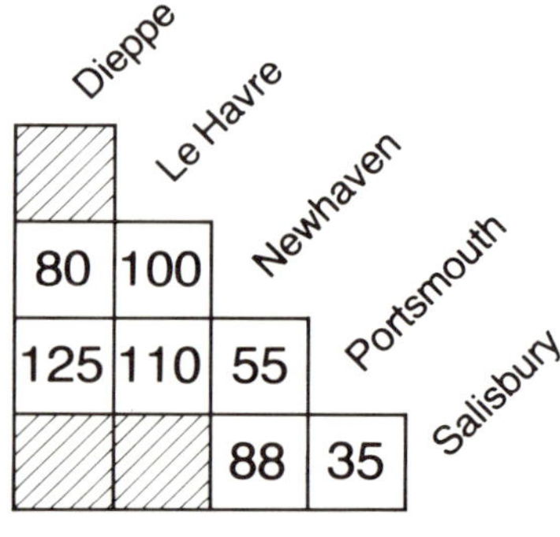

Distance in miles

	Dieppe	Le Havre	Newhaven	Portsmouth
Le Havre				
Newhaven	80	100		
Portsmouth	125	110	55	
Salisbury			88	35

All distances are approximate

Figure 2

Le Havre Super Day £12.50

Two Adult passengers for the price of one until 20th May.

This trip gives you more time ashore to enjoy all of Le Havre's delights.

Sail out from Portsmouth 0830
Return sailing from Le Havre 2330 same day
SEVEN HOURS ASHORE

Price any day	Adult £12.50	Child £7.00†

*Free litre of whisky or gin for all persons over 17 years of age during period 1 April to 20th May inclusive, and is part of their Duty Free allowance.

†Child fares: Refer to children 4 to under 14 years of age. Children under 4 travel FREE

....and Bikes free!

NEWHAVEN-DIEPPE

PRICES HELD

SAME DAY EXCURSIONS TO DIEPPE
FROM £9.00 PER PERSON

TIMETABLE

Passengers should arrive at Newhaven at least 30 minutes before advertised sailing times

NEWHAVEN	dep.	0700	1000		
DIEPPE	arr.	1100	1400		
DIEPPE	dep.	0400	1245	1630	0015
NEWHAVEN	arr.	0800	1645	2030	0445

Day Excursions: The return journey must be made on the day of outward travel or on the 0015 hrs sailing the following day.

A Day Out

ON YOUR OWN

Look at the graph in Figure 1.

1 How many journeys does it show?

2 How long does the journey from Manchester to Blackpool take on this day?

3 What time is it at point X?

4 What happens at 9:51 a.m.?

5 What happens to the bus from Blackpool at 9:30 a.m.?

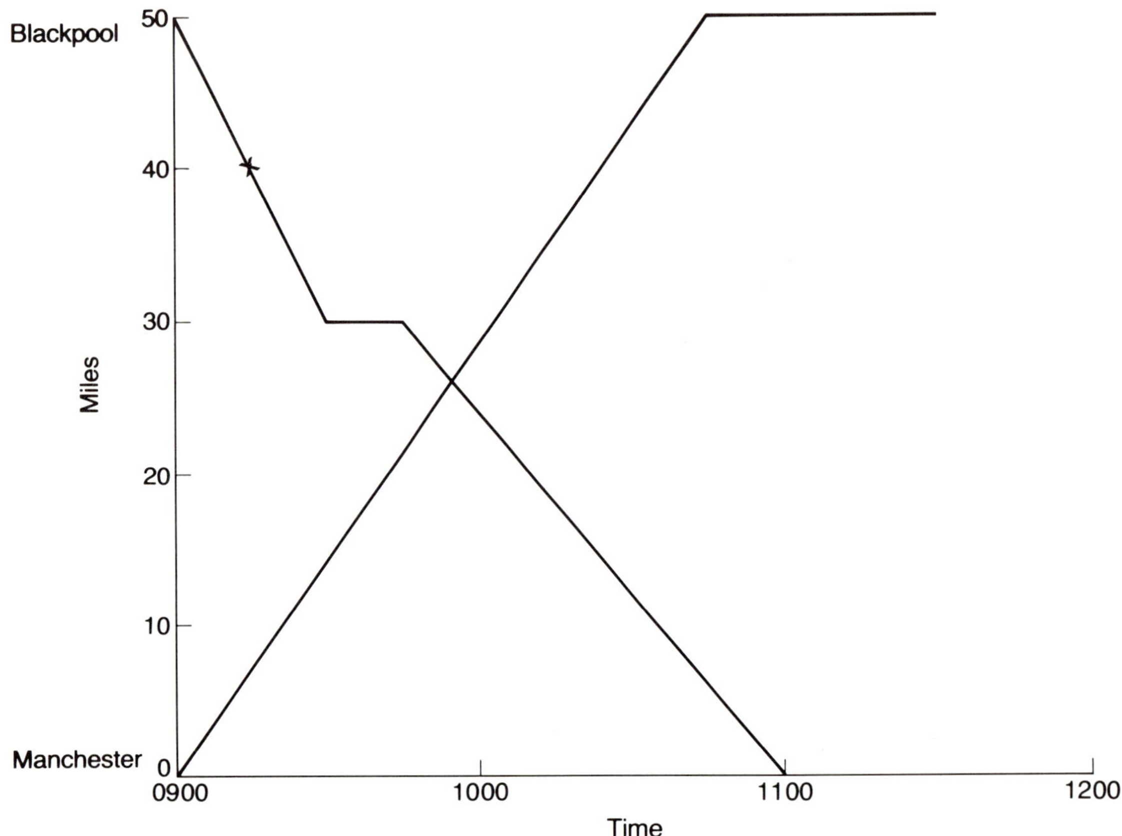

Figure 1

Double-Glazing

Topic	Practical geometry.
Content	Measuring. Area of rectangle. Metric length/area conversion. Scale drawing. Costing. Extraction of information.
Assumed knowledge	Metric length relationships. Use of tape-measure. Calculation of area of rectangle.
Integration	UM3: to precede *Area: parallelogram.* SUM3: to precede *Perimeters and areas: rectangles.*
Administration	Small groups. Metric tape-measures will be needed. Discuss which windows would be sensible for step 3. Less able pupils may need quite a lot of help.

Answers and Marking Guide

Step	Notes	Max. Marks
1	Frame packs 1050 mm and 1500 mm	2
	$1{\cdot}05\,\text{m}\times1{\cdot}5\,\text{m}\simeq1{\cdot}57\,\text{m}^2$	2
2	(a) 11 @1050 → 11×£5·26 =£57·86	2
	3 @1500 → 3×£7·49 =£22·43	2
	Total =£80·29	1
	(b) 4 @1 m^2= 4 m^2	1
	3 @1·5 m^2=4·5 m^2	1
	8·5 m^2 @£10·80=£91·80	2
3	Sketch and measurements	2
	(a) Correct packs	2
	(b) Correct amount of glazing	2
4	Clear order	1
	Correct costing	2
5	Good designs and adverts	3
		25

Suggested Grading Scheme

	A	B	C	D	E
Level 3	23+	20–22	17–19	14–16	12–13
Level 2	18+	16–17	13–15	11–12	9–10
Level 1	15+	13–14	11–12	9–10	7–8

On Your Own Answers

		Max. Marks
1	17·50	3
2	2·16 m^2	3
3	Check with tracing or OHP (see Figure T1)	4
		10

Scale: 1 cm to 1 m

Figure T1

Double-Glazing

1 Look at the window in Figure 1. Using the information on the Worksheet what size Quickglaze frame packs will you need to buy? Remember to buy one pack for the top/bottom and one pack for the sides. Now change the height and width measurements to metres and find the area of glazing material you will need to buy.

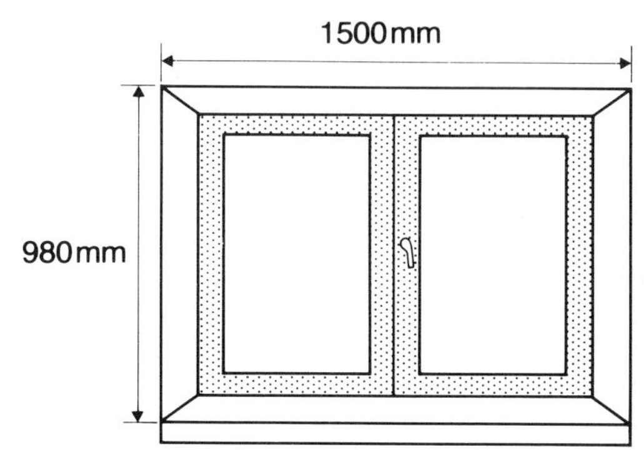

Figure 1

2 Look at Figure 2. It shows a plan of the windows at the front and back of a house.

(a) How much would the frames cost for these seven windows?

(b) How much would plastic sheet cost for these seven windows?

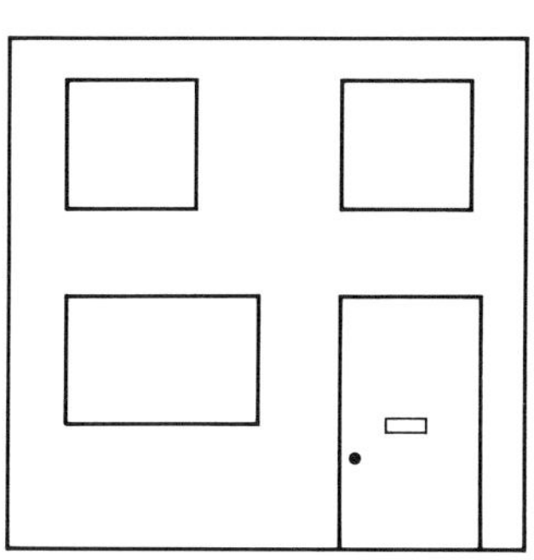

Figure 2

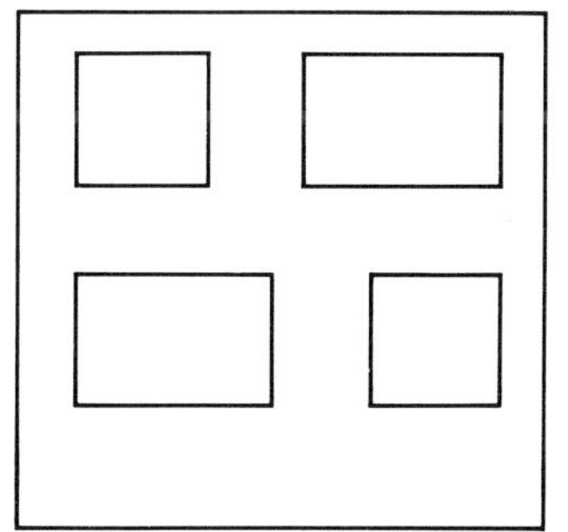

Scale: 1 centimetre = 1 metre

3 Measure a window in your classroom (or some other suitable window). Measure in millimetres, following the instructions 'How to measure your windows' on the Worksheet. Sketch the window, marking in the measurements.

(a) Following the instructions work out which two Quickglaze frame packs you will need to buy.

(b) Now work out how much glazing material you will need.

4 Write an order for the materials you need for your classroom window. Work out the total cost of frame packs and 3 mm glass.

5 Gardeners use cloches to protect seedlings when they are small and vulnerable during the early part of the year. Design your own cloche. You could make a model of it. Work out some suitable sizes and then work out the cost of plastic sheet (still at £10·80 per square metre). Allow 25% extra for the frame.

Draw an advert to put in a gardening magazine to advertise your cloche/cloches.

QUICKGLAZE

FRAMES

**Each window needs one frame pack for the sides and one for the top/bottom.
This system is not suitable for windows which are wider or longer than 1950 mm.**

		OUR PRICE
FRAME	600mm	£3.03
	750mm	£3.87
	900mm	£4.48
	1050mm	£5.26
	1200mm	£5.96
	1350mm	£6.69
	1500mm	£7.49
	1650mm	£8.25
	1800mm	£8.75
	1950mm	£9.47

All kits include fitted draught-proofing seals and step-by-step instructions, together with accessories and fitting aids such as screws and wallplugs.

CHOICE OF GLAZING MATERIALS

PLASTIC SHEET £10.80 per sq. metre
GLASS (3mm thickness) £13.50 per sq. metre.

HOW TO MEASURE YOUR WINDOWS

Measure the width and the height of the gap which the secondary glazing frame has to fit into.

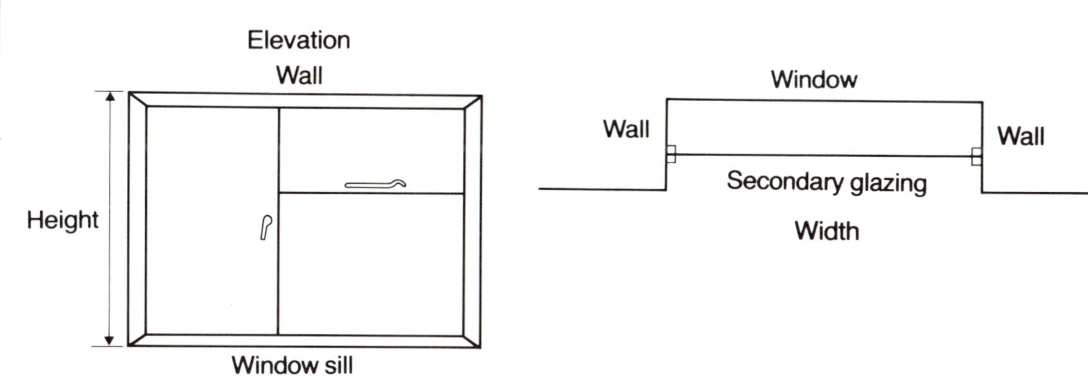

Double-Glazing

ON YOUR OWN

1 Glass costs £14·00 per square metre. How much will a sheet of glass measuring 1·25 square metres cost?

2 What will be the area in square metres of a piece of glass that measures 1800 mm by 1200 mm?

3 Using a scale of 1 centimetre to 1 metre, draw as accurately as you can a room measuring 3·5 metres by 5·5 metres.

Get Fit!

Topic	Statistics.
Content	Substitution in formulae. Statistical averages. Frequency tables. Frequency curves.
Assumed knowledge	Drawing frequency curves. Calculating mean, median and mode from frequency tables.
Integration	UM3: to follow *Statistics: frequency tables.* SUM3: to follow *Averages.*
Administration	Pairs or small groups. It is best if pupils take each others' pulse rates. Any energetic exercise will do instead of star jumps.

Answers and Marking Guide

Step	Notes	Max. Marks
3	Correct use of the formula	3
4/5	Accurate and neat curve. See Figure T1	3
	Mean = 42 780 / 600 ≃ 71·3	2
	Mode is 70	2
	Median is 70	2
	Say: above 80 fit 60–80 average under 60 unfit	3
6	Presentation of results	3
	Correct curve and answers	3
7	Ideas and reasoning	2
	Realistic test	2
		25

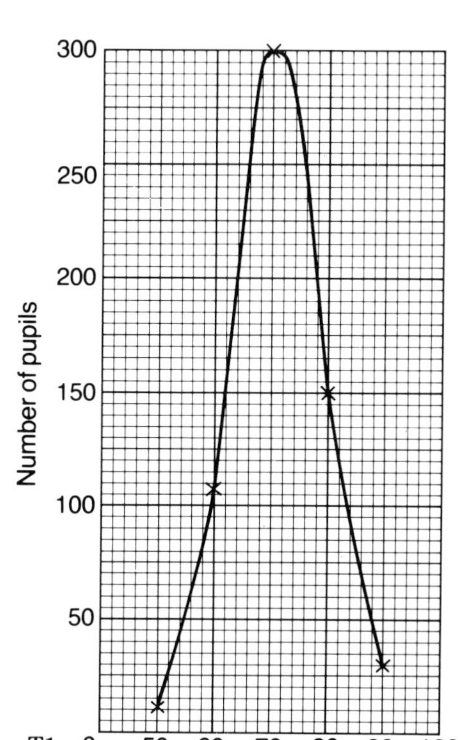

Figure T1

Suggested Grading Scheme

	A	B	C	D	E
Level 3	21+	18–20	15–17	12–14	10–11
Level 2	16+	14–15	11–13	9–10	7–8
Level 1	12+	10–11	8–9	6–7	4–5

On Your Own Answers

		Max. Marks
1	72·7 (accept 73, or 70 if rounded to nearest 10)	2
2	No, the time must be the same	2
3	Mean ≃ 75·3	2
	Median is 80	2
	Mode is 90	2
		10

Get Fit!

There are many tests for measuring physical fitness. Some of these tests are based on the pulse rate measured before and after exercising (Figure 1). The average pulse rate at rest is usually said to be between 60 and 80 beats per minute, but things such as injury, illness, emotion and exercise can all increase the pulse rate. The length of time the pulse rate takes to return to normal after exercising will give you a guide as to how fit a person is.

1 What is your pulse rate at rest?

2 Try the exercise shown in Figure 2 for 90 seconds, then sit down quietly for 1 minute. Then record your pulse rate.

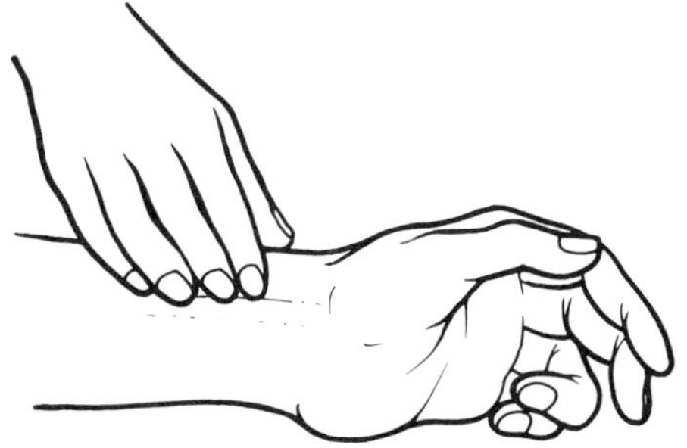

Pulse beats occur when the contraction of the heart thrusts blood through the arteries making them expand. The pulse may be taken at many places but it is normally taken at the wrist about 1 centimetre in from the thumb side of the arm.

Count the number of beats which occur in 30 seconds and multiply by 2 for the rate per minute.

Figure 1

Star jumps
Start with your feet together and your hands at your side. After each jump your hands should return to your sides and your feet should be together.

Figure 2

3 To work out how fit a person is, the following can be used as a rough guide:
Multiply the number of seconds spent doing the exercise by 100.
Divide your answer by your pulse rate (per minute, after exercise) multiplied by 1·5.
Round your answer to the nearest 10.

$$\frac{90 \times 100}{\text{Pulse rate} \times 1{\cdot}5}$$

4 In a group of 600 pupils the following results (rounded to the nearest 10) were gained using the formula in step 3:

Fitness score	**Number of pupils**
50	12
60	108
70	300
80	150
90	30

Which group does your score fit into?

5 Draw a frequency curve for the information in step 4. What are the mean, median and mode for these statistics? Can you draw any conclusion from your findings?

6 Collect information from the rest of your class, present and comment on the information.

7 Do you think the star jumps exercise is a suitable test of fitness? Design some tests of your own – they could be for fitness, strength, stamina etc. Test them out and comment on them. Are there other factors which affect your test results?

Get Fit!

ON YOUR OWN

1 Stephen has a pulse rate of 110 after strenuous exercise which lasted 2 minutes. If you use the formula:

$$\text{Score} = \frac{T \times 100}{1{\cdot}5 \times P},$$

where T = time spent exercising in seconds, and P = pulse rate per minute, what will his score be?

Using a calculator

T [×] 100 [÷] 1·5 [÷] P =

2 After doing the same exercise for 1 minute, Jayne has an index of 92. Is it fair to say that Jayne is fitter than Stephen?

3 From this class set of 30 results, what are the mean, median and mode?

Score	Frequency
50	3
60	4
70	7
80	7
90	8
100	1

Cheese Snacks

Topic	Problem-solving.
Content	Knowledge of the cylinder. Meaning of circumference/radius/perimeter. Volume of a cylinder.
Assumed knowledge	Formula for volume of a cylinder. Finding a radius from a circumference.
Integration	UM3: to follow *Volume of a cylinder.* SUM3: to follow *Circles: circumference.*
Administration	Group or individual. Many will make and measure the cylinders, using sellotape or paper clips, to hold them together. SUM3 pupils will need to be told the formula $V=\pi r^2 h$ for the volume of a cylinder, with suitable discussion.

Answers and Marking Guide

Step	Notes	Max. Marks
1	Look for any methodical approach. Pupils should realise that it is volume they are looking for	5
	The largest volume is 505·15 cm^3. See Figure T1	5
	The radius of the plastic stopper is 3·66 cm	2
2	Labels are 24 cm long. The roll should be 6 m (25 labels)	3
		15

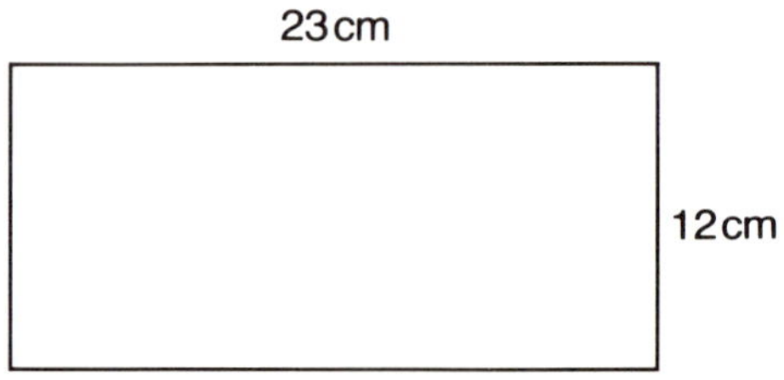

23 cm

12 cm

Figure T1

Suggested Grading Scheme

	A	B	C	D	E
Level 3	14+	12–13	10–11	8–9	6–7
Level 2	12+	10–11	8–9	6–7	4–5
Level 1	10+	8–9	6–7	4–5	2–3

On Your Own Answers

		Max. Marks
1	Perimeter: 140 cm	3
2	318·31 cm^3	3
		6

Cheese Snacks

A company that manufactures crisps and snacks decide that they want to add a new product to their range, small cheese biscuits.

These biscuits will be put in cylindrical cartons made of card with a plastic stopper at both ends. See Figure 1.

They work out that they can use a piece of rectangular card for the carton which has a perimeter of 70 centimetres.

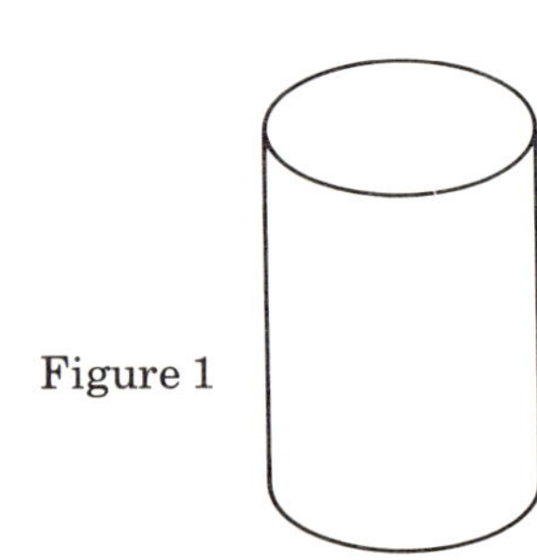

Figure 1

Making the carton

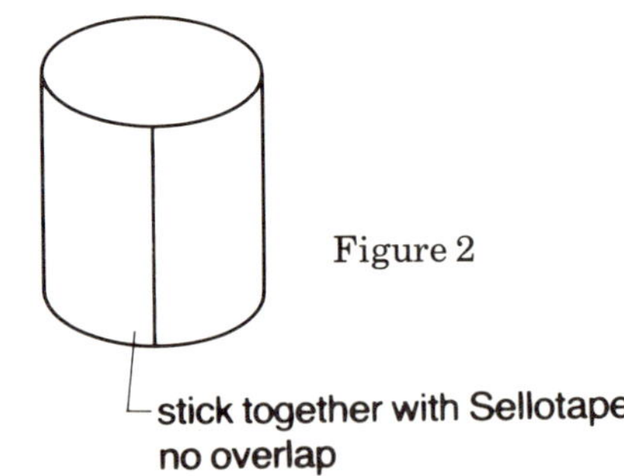

Figure 2

1 What exact size should they make the carton if it is to contain as many biscuits as possible, and what size plastic stoppers should they have made?

2 Labels for the boxes of cheese biscuits are produced on a machine which leaves them all on a roll of paper. The labels are one centimetre longer than the circumference of the boxes. If the roll of paper labels is to be in whole metres, what length should the roll be in order for there to be no waste?

Cheese Snacks

ON YOUR OWN

1 A cardboard cylinder is made with a height of 50 centimetres and a circumference of 20 centimetres. If there is no overlap at the join, what must be the perimeter of the rectangle of card used to make the cylinder?

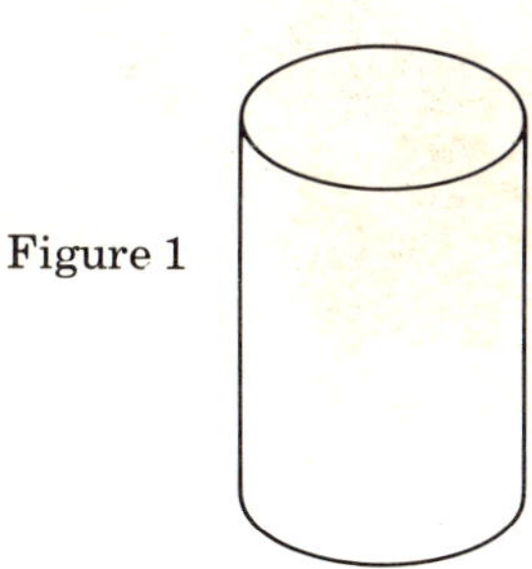
Figure 1

2 Find the volume of the cylinder in Figure 2. Remember that the volume of a cylinder is $\pi r^2 h$.

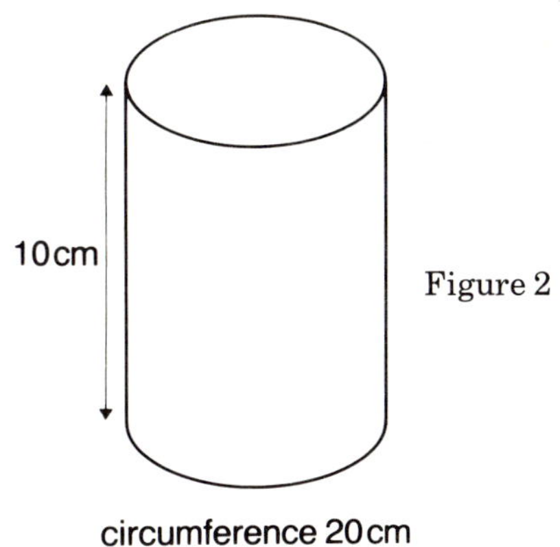

Figure 2

Mirror, Mirror

Topic	Investigation.
Content	Symmetry. Congruence. Organisation of an investigation.
Assumed knowledge	None.
Integration	UM3 and SUM3: to precede *Symmetry*.
Administration	Pairs or small groups. Encourage a logical, organised approach. With less able pupils point out that 'one line of symmetry' means *just* one line. Pupils should be urged to 'keep looking' for more patterns in steps 2, 4, 6, 7 and 8.

Answers and Marking Guide

Step	Notes	Max. Marks
1	Following the instructions	1
	Writing about its symmetry	1
2	See Figure T1	
	Copy with symmetry line drawn	1
	Correct square shaded	1
	Correct alternative	1
3	Two	2
4	See Figures T2 and T3	
	One square shaded	1
	Other ways	2
5	(a) and (c)	1

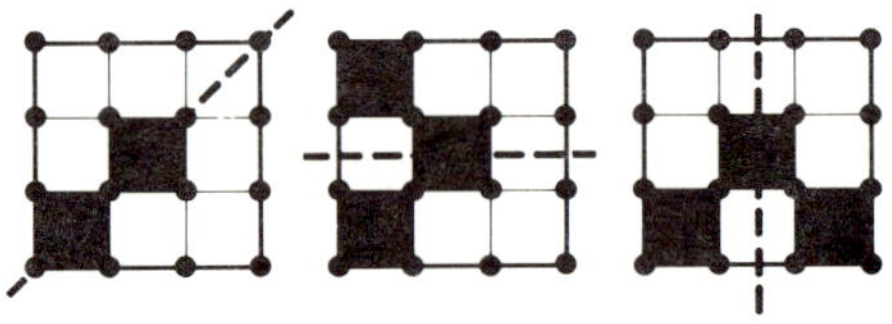

Figure T1

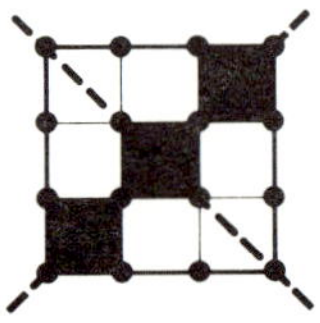

Figure T2

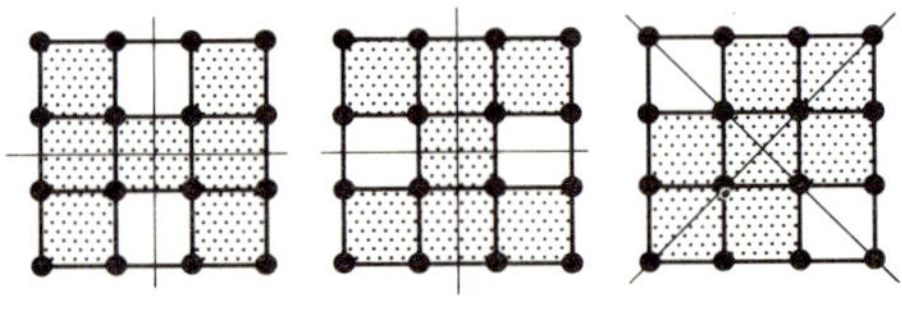

Figure T3

		Max. Marks
6	Thirteen ways. See Figure T4	
	One way 1 mark	
	Up to eight ways 2 marks	
	Up to twelve ways 3 marks	
	All thirteen ways 4 marks	4

Figure T4

7	Three parts, see Figure T5	1
	Four parts, see Figure T6	3

Figure T5

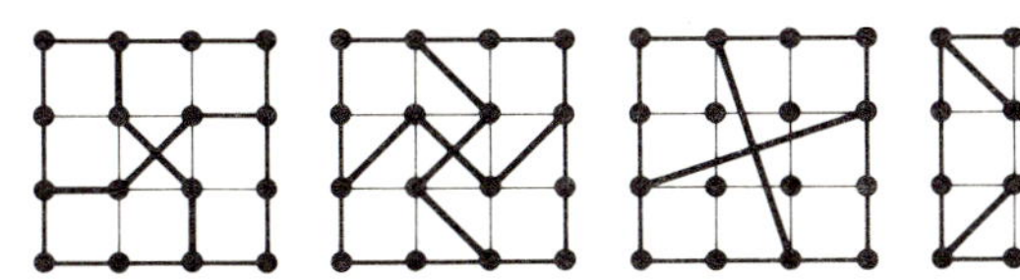

Figure T6

8	See Figures T7–T10	
	Evidence of organisation of method	2
	At least four ways 1 mark	
	Up to ten ways 2 marks	
	Up to twenty ways 3 marks	
	Over thirty ways 4 marks	4
9	Further investigation	5
		30

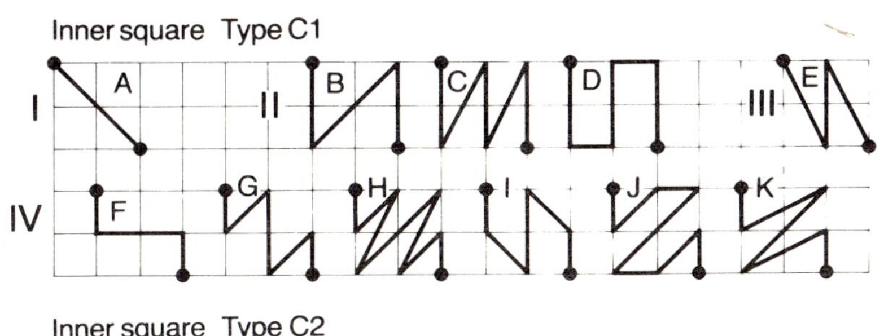

Figure T7

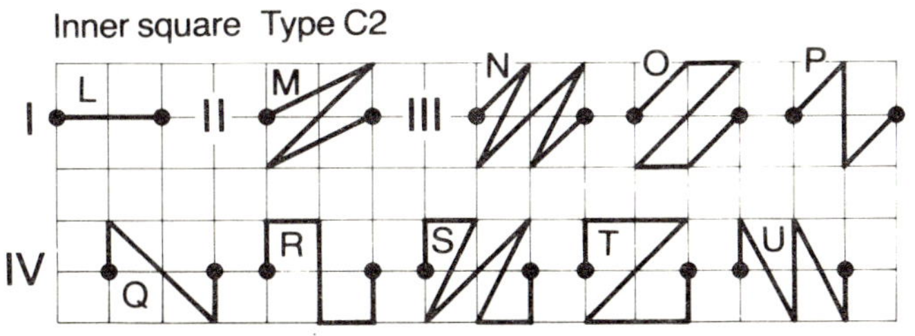

Figure T8

Outer ring starts for Type C1

1 A to K (11)
2 A to K (11)
3 A to K (11)
4 A to K (11)
5 A to K (11)
6 none (duplicates C1/2 and C2/4) (0)
7 A and F (2)
8 A (1)

Total 58

Figure T9

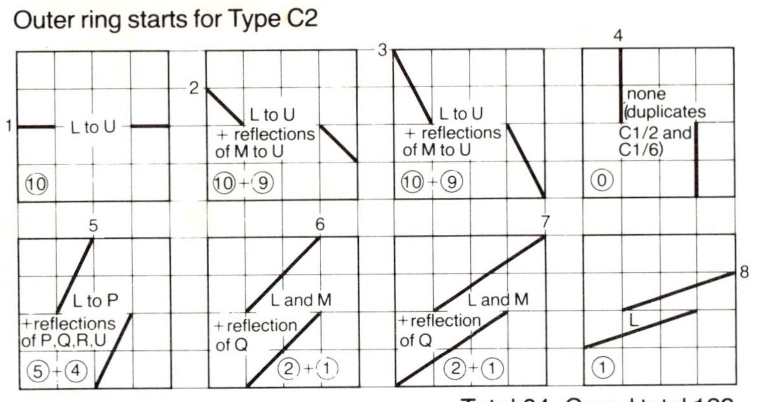

Figure T10

Suggested Grading Scheme

	A	B	C	D	E
Level 3	24+	22–24	17–22	14–17	13
Level 2	19+	16–19	14–16	12–14	11
Level 1	15+	13–15	11–13	9–11	8

On Your Own Answers

		Max. Marks
1	(a), (d) and (e) are congruent	3
2	AEC and BFC	2
3	The same size and shape	3
		8

Mirror, Mirror

1 Take a rectangular sheet of paper, fold it in half. Draw a shape on your piece of paper as shown in Figure 1. You can use any shape you like. Cut along the line you have drawn. Open out your paper. What do you notice about the shape you have cut out?

The fold line on your paper is called the line of symmetry.

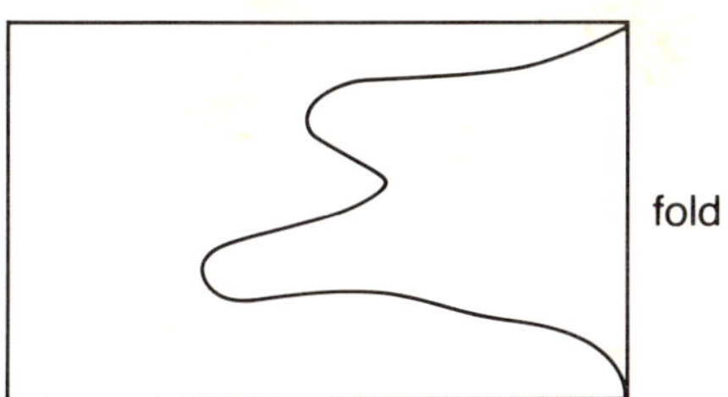

Figure 1

2 The pattern in Figure 2 has one line of symmetry. Copy Figure 2 onto Worksheet One and draw the line of symmetry. Now copy Figure 2 again, then shade one more square so that the whole figure still has just one line of symmetry.

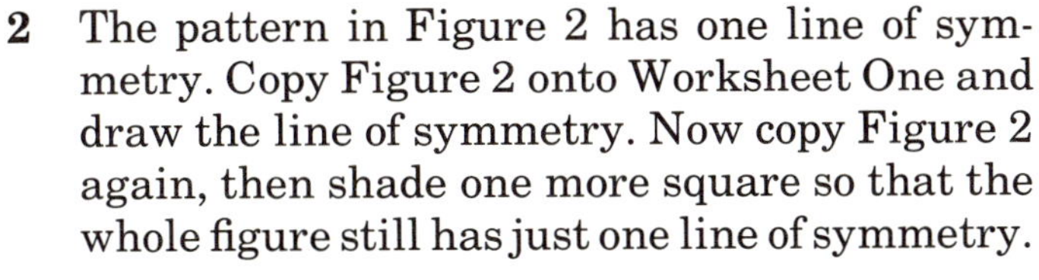

Can you find any other way of shading one extra square to give Figure 2 just one line of symmetry?

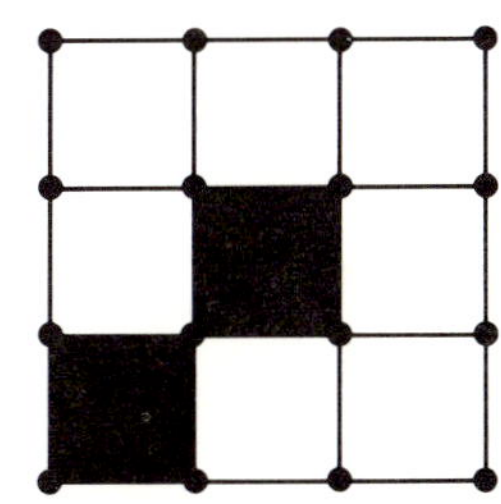
Figure 2

3 If you fold a rectangular piece of paper as in Figure 3 and then cut out a shape, how many lines of symmetry will your shape have when you unfold it?

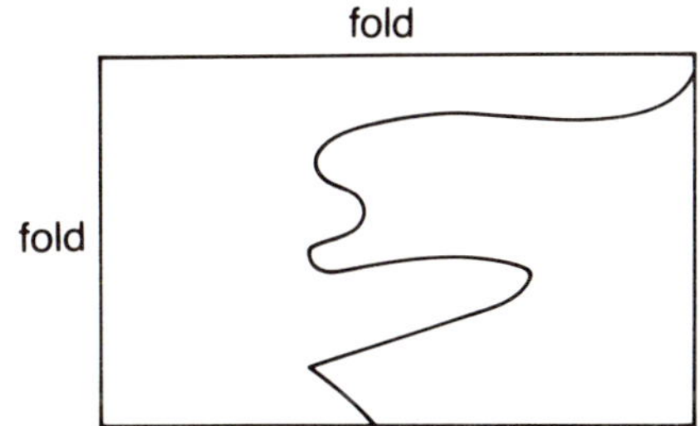

Figure 3

4 Copy Figure 2 again. This time, shade just one more square to give the whole figure two lines of symmetry. Are there any other ways of doing this?

If you could shade more than one extra square can you find any other ways of giving the figure two lines of symmetry?

5 Congruent means the same size and the same shape. Which of the shapes in Figure 4 are congruent?

6 Find as many ways as you can to divide the squares drawn on Worksheet One into two congruent parts. All straight lines must join two dots. A method is only different if it gives a shape that is not congruent to any other one that you have already drawn. Try to organise your ways into sets of the same type.

7 How many ways can you find to divide the squares on Worksheet One into three congruent parts, again only using straight lines that join dots? What about four congruent parts?

8 Now use Worksheet Two. How many ways can you find to divide these squares into two congruent parts? To be sure that you have found them all you will need to organise your methods carefully. Explain clearly how you do this.

9 Investigate further.

(a)

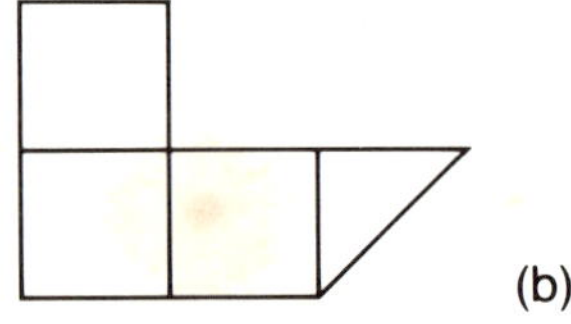
(b)

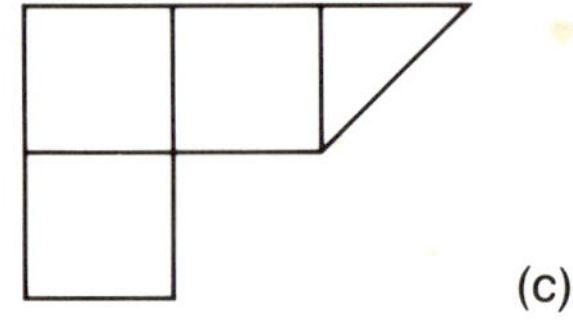
(c)

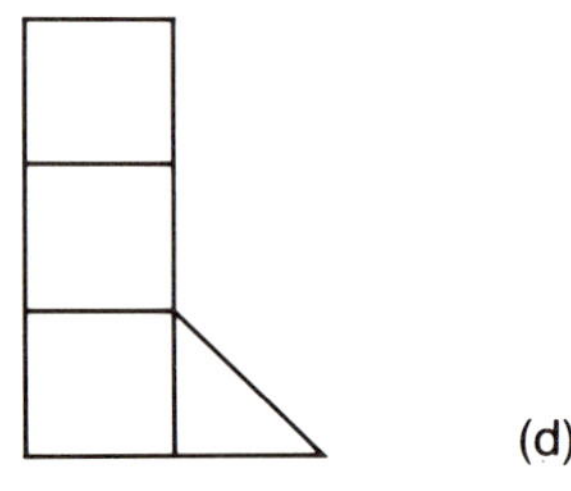
(d)

Figure 4

Mirror, Mirror

WORKSHEET ONE

Mirror, Mirror WORKSHEET TWO

Mirror, Mirror

ON YOUR OWN

1 Which of the shapes in Figure 1 are congruent?

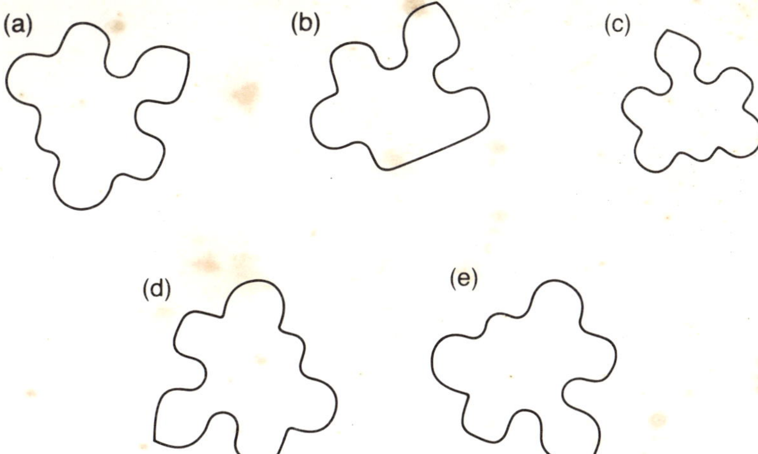

Figure 1

2 In Figure 2 find as many pairs of congruent triangles as you can.

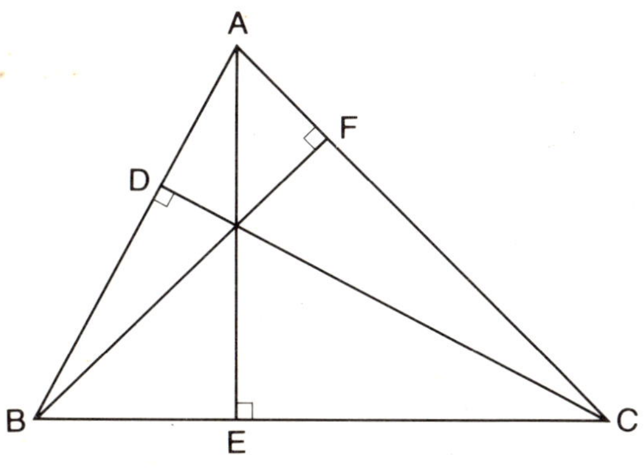

Figure 2

3 Describe in your own words what congruent means.